Introduction 1

Part I: Theory of Linear Noisy Networks

Noise Characterization of Linear Circuits, *K. Hartmann* (*IEEE Transactions on Circuits and Systems*, October 1976) 3

Available Power Gain, Noise Figure, and Noise Measure of Two-Ports and Their Graphical Representations, *H. Fukui* (*IEEE Transactions on Circuit Theory*, June 1966) 13

Part II: Noise in Electronic Amplifier Circuits

The Design of Low-Noise Amplifiers, *Y. Netzer* (*Proceedings of the IEEE*, June 1981; *Comments*, August 1982) 19

Amplifier Techniques for Combining Low Noise, Precision, and High-Speed Performance, *G. Erdi* (*IEEE Journal of Solid-State Circuits*, December 1981) 33

Low-Frequency Noise Considerations for MOS Amplifiers Design, *J-C. Bertails* (*IEEE Journal of Solid-State Circuits*, August 1979) 42

Part III: Computer-Aided Noise Analysis

Computationally Efficient Electronic-Circuit Noise Calculations, *R. Rohrer, L. Nagel, R. Meyer, and L. Weber* (*IEEE Journal of Solid-State Circuits*, August 1971) 45

Network Sensitivity and Noise Analysis Simplified, *F. H. Branin, Jr.* (*IEEE Transactions on Circuit Theory*, May 1973) 55

An Efficient Method for Computer Aided Noise Analysis of Linear Amplifier Networks, *H. Hillbrand and P. H. Russer* (*IEEE Transactions on Circuits and Systems*, April 1976) 59

Part IV: Noise in Active Filters

Noise Performance of Low Sensitivity Active Filters, *L. T. Bruton, F. N. Trofimenkoff, and D. H. Treleaven* (*IEEE Journal of Solid-State Circuits*, February 1973) 63

Noise Performance of *RC*-Active Quadratic Filter Sections, *F. N. Trofimenkoff, D. H. Treleaven, and L. T. Bruton* (*IEEE Transactions on Circuit Theory*, September 1973) 70

Noise Performance Limitations of Single Amplifier *RC* Active Filters, *J. W. Haslett* (*IEEE Transactions on Circuits and Systems*, September 1975) 79

Part V: Noise in Switched-Capacitor Filters

Noise Sources and Calculation Techniques of Switched Capacitor Filters, *J. H. Fischer* (*IEEE Journal of Solid-State Circuits*, August 1982) 85

Noise Analysis of Switched Capacitor Networks, *C-A. Gobet and A. Knob* (*IEEE Transactions on Circuits and Systems*, January 1983) 96

Author Index 103

Editor's Biography 105

Advances in Circuits and Systems

SELECTED PAPERS ON
NOISE IN CIRCUITS AND SYSTEMS

EDITED

M

Profes
Univer

A SER
IEEE

PRESS

The Institute of Electrical and Electronics Engineers, Inc., New York

IEEE PRESS
1988 Editorial Board
R. F. Cotellessa, *Editor in Chief*
J. K. Aggarwal, *Editor, Selected Reprint Series*
Glen Wade, *Editor, Special Issue Series*

James Aylor	W. K. Jenkins	A. C. Schell
F. S. Barnes	A. E. Joel, Jr.	L. G. Shaw
J. E. Brittain	Shlomo Karni	M. I. Skolnik
B. D. Carrol	R. W. Lucky	P. W. Smith
Aileen Cavanagh	R. G. Meyer	M. A. Soderstrand
D. G. Childers	Seinosuke Narita	M. E. Van Valkenburg
H. W. Colborn	J. D. Ryder	John Zaborsky
J. F. Hayes		

W. R. Crone, *Managing Editor*
Hans P. Leander, *Technical Editor*
Laura J. Kelly, *Administrative Assistant*

Allen Appel, *Associate Editor*

Copyright © 1988 by
THE INSTITUTE OF ELECTRICAL AND ELECTRONICS ENGINEERS, INC.
345 East 47th Street, New York, NY 10017-2394
All rights reserved.

PRINTED IN THE UNITED STATES OF AMERICA

IEEE Order Number: PP0230-3

Library of Congress Cataloging-in-Publication Data

Selected papers on noise in circuits and systems.

(Advances in circuits and systems)
"A series published for the IEEE Circuits and Systems Society."
Includes index.
1. Electronic circuits—Noise. I. Gupta, Madhu S.
II. IEEE Circuits and Systems Society. III. Series.
TK7867.5.S36 1988 621.3815'3 87-32427

ISBN 0-87942-239-4 (pbk.)

Introduction

This short volume is a compilation of thirteen papers on the subject of noise in electrical circuits and systems. Its purpose is to bring together some of the most significant papers on this subject that have previously appeared in the various journals published or sponsored by the IEEE Circuits and Systems Society. Such a collection will hopefully be useful both to professionals who work in the field, and to students and other beginners whose work requires a familiarity with this field.

It is apparent that a compilation of research papers is essentially a substitute for a comprehensive, authoritative, and up-to-date review article. But the amount of time and the level of expertise required to write such a review article are such that few authors can be coaxed to write them. Hence the need for a collection such as this. Admittedly, a compilation of reprinted research papers lacks the uniformity and tutorial background that a review article would have; a partial compensation for that is the fact that the original papers provide a first-hand account of research at the frontier of the field.

Two of the obvious advantages of a reprint collection such as this are the convenience for the reader, and the wider dissemination of the literature among those interested in the field. But there are several additional benefits of a compilation of this kind that are more significant. For students and new researchers who are looking for interesting fields or problems, a compilation provides a recapitulation of an entire field of study, and is therefore useful for identifying the areas that are professionally active and attractive to research workers, and that have been a fertile source of research problems. For those who are looking for answers rather than problems in the field, the reprint volume consolidates the knowledge in the field of specialization with which the papers deal, and it thus defines the state of the art reached by the professionals working in that field, the class of problems that are known to be solved or solvable, and the boundary of knowledge in the field. Finally, for all readers, the compilation serves as a guide to some of the significant reading material in the field.

The subject of noise in circuits and systems is a large one. The scope of the present volume is limited to a small part of the subject that lies within the domain of interest of the IEEE Circuits and Systems Society. This part deals with the network representation of the noise generated in electronic devices and circuit elements, the determination therefrom of the noise performance of the overall network, and the circuit design techniques for minimizing the effects of noise. Excluded is the "noise" literature in which the term noise is used to describe "error" (e.g., quantization noise or roundoff noise, as in digital filters), or in which the term noise has become traditional in place of "variability" (e.g., noise margin or noise immunity, as in switching circuits).

The study of noise in circuits can be said to have seriously begun only in the second half of the twentieth century. Physicists began studying noise in the form of Brownian noise at the beginning of this century, and gained some understanding of the noise in such individual circuit elements as resistors and vacuum tubes by the 1930s. In the early days of electronics, the ability of a circuit to process signals was limited by man-made and natural interference, and there was no motivation for the study of circuit noise. It was not until the late 1930s, when circuits at "short-wave" frequencies were designed, that the need arose for characterizing and reducing the noise generated within the circuit. Beginning with the early 1940s, attention was focused on the description of the sensitivity of radio receivers, and the mathematical methods of analyzing random signals received attention during and after the Second World War. The birth of the transistor gave a further impetus to research on noise starting with the early 1950s. The first systematic study of noise in circuits from the network theoretic viewpoint was that of Rothe and Dahlke in the 1950s, which dealt with the representation of noise in one of the simplest classes of networks: the linear two-ports. Since that time, a significant number of the major papers on the subject of noise in circuits have appeared in the journals of the IEEE Circuits and Systems Society. The selection of papers for this volume was limited to these journals.

The class of circuits for which noise has been studied is very large and, as one would expect, it has changed with time. For example, the literature of the 1950s and 1960s contains numerous papers on the study of noise in transistor circuits, in periodically time varying (parametric) circuits, and in circuits containing negative-resistance devices. None of these subjects is represented in this compilation, as they might have been if the compilation had been made at an earlier time. The present volume emphasizes the basic theory of noisy networks, as well as topics of more current interest. The thirteen papers reprinted here are grouped into five subject categories: Theory of Linear Noisy Networks, Noise in Electronic Amplifier Circuits, Computer-Aided Noise Analysis, Noise in Active Filters, and Noise in Switched-Capacitor Filters. The scope of the reprinted papers in each of these areas is briefly described in the following.

A large amount of literature has appeared on the subject of noise in linear networks, dealing with the various forms of representation of noise in circuits; generalizations to negative resistances, nonreciprocal networks, n-ports, etc.; calculation of parameters characterizing the network noise performance; and their interrelationships. A comprehensive survey and a sizable bibliography of this literature appear in the first reprinted paper by Hartmann. The second paper by Fukui is based on the Rothe-Dahlke representation of noise in a linear two-port. It introduces several parameters useful for the characterization of linear twoport noise, deduces expressions for these parameters from the Rothe-Dahlke model, and describes a graphical method of representation for the results.

While the low-noise circuit design techniques have been applied to a variety of circuits, they are most developed for amplifier circuits. A survey of the design considerations in low-noise transistor circuit design is contained in the paper by Netzer. The next two papers deal with two specific types of amplifiers where noise reduction is of great practical importance. The paper by Erdi describes a technique for low-noise design of operational amplifiers. MOS transistors generate a large amount of low-frequency noise, and the design of low-noise MOS transistor circuits requires special attention, as discussed in the paper by Bertails.

Once an equivalent circuit representation for the noise generated in a circuit has been established, the next problem is that of computing the effect of internal noise sources of the network on its response. The increase in circuit complexity, and the development of integrated circuits, motivated the search for efficient methods of computer-aided noise analysis beginning in the 1960s. Three papers on this subject are included in this volume. In the first paper by Rohrer et al., the noise sources in the noise equivalent circuit representation of a network are assumed to be uncorrelated with each other, and the method of adjoint networks is used to determine the noise output of the network. The second paper by Branin retains the assumption of uncorrelated noise sources, but does not employ the concept of adjoint network for response calculation. The third paper, by Hillbrand and Russer, is applicable only to those twoport networks which can be decomposed into series, parallel, and cascade combinations of more elementary two-ports, and expresses the noise of each elementary two-port in terms of its correlation matrix. While limited to a class of networks, this method does not require the assumption of uncorrelated noise sources for individual elementary two-ports.

The next three papers deal with the noise performance of one class of circuits which are common building blocks, and are amenable to systematic noise analysis: the active filters. The first paper by Bruton et al. determines the noise performance of two basic active filter structures, and then deduces the theoretical noise limitation of these filter realizations. The second paper by Trofimenkoff et al. calculates the noise performance of a quadratic filter section, assuming that the operational amplifier used is ideal in all respects other than noise. The third paper, by Haslett, analyzes noise in active filters employing Norton quad amplifiers, and compares their noise performance with that of filters employing the conventional differential voltage amplifiers.

The last two papers reprinted in this volume are devoted to the calculation of noise in switched-capacitor filters. The paper by Fischer contains a detailed description of the basic technique of noise performance calculation in switched-capacitor filters. The final paper by Gobet and Knob describes the construction of complete noise models for switched-capacitor filters, and the calculation of the noise response from the models. Both papers present experimental data in support of the results.

July 13, 1987

Madhu S. Gupta
University of Illinois at Chicago
Chicago, Illinois

Part I: Theory of Linear Noisy Networks

Noise Characterization of Linear Circuits

KARL HARTMANN, MEMBER, IEEE

Invited Paper

Abstract—The noise characterization of linear circuits is discussed in connection with the transformation of the noise parameters. The noise performance of two-ports is treated, as are methods for linear devices including correlated, uncorrelated, and partially correlated noise sources. Finally, the experimental determination of the noise parameters is described. Special attention is given to negative source resistances. It is shown that the noise performance of any linear two-port network can be completely characterized by 12 parameters; four noise and four complex network parameters.

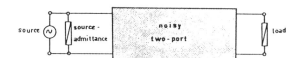

Fig. 1. Noise performance of single noisy two-port connected to external source with source admittance can be characterized by four noise parameters.

I. INTRODUCTION

ANY UNWANTED signal in a transmission system is called noise. In some cases noise can be a useful signal for the experimental determination of physical quantities. Unwanted signals can be caused by a number of sources. In linear circuits, the primary cause is the quantum nature of the electrical charge flowing in electrical devices. Physicists made the first investigations of noise. In 1827 Brown [1] observed fluctuations about the equilibrium state of a physical system. Not until the beginning of this century was this phenomenon analyzed extensively. Communication engineers applied the early results of Einstein [2], and others [3]–[5], to the solution of noise problems by introducing electrically defined noise quantities. The basic concepts of the noise figure was introduced by Burgess [6], North [7], and Friis [8] in the 1940's. The basic definitions were then extended by other authors. Goldberg [9] has shown a possibility of noise factor reduction by mismatching impedances. Haus and Adler [10] included analysis of receivers with a negative resistance. Since the noise figure definition is only a man-made definition rather than a quantity deduced from clearly defined postulates or laws of nature, it may be assumed that in practice not all requirements are satisfied using this definition. For this reason Harris [11], McCurley and Blake [12], and Norton [13] worked out other definitions which are worth discussing at greater length. But this is beyond the scope of this paper. The present contribution is based on the noise figure definition of the IEEE standards [14]–[16]. The noise performance of two-ports and their combinations are treated along these lines. The theory forwarded can be expanded to n-ports. The results obtained can serve as a design tool for constructing low-noise circuits. In constructing these circuits, use is made of the studies on the noise behavior of electron devices which were carried out by Strutt [17], van der Ziel [18], and Bennet [19].

II. NOISE CHARACTERIZATION OF LINEAR CIRCUITS

The noise behavior of a single linear noisy two-port connected to an external source with a source admittance (Fig. 1) can be characterized by *four parameters*, namely, by two *power spectral densities* and the *real* and *imaginary part* of the corresponding *cross spectral density*. This number of parameters must be increased for more complicated two-port configurations. The power spectral density is defined as the Fourier transform of the well-known autocorrelation function; the cross spectral density is defined as the Fourier transform of the cross-correlation function [20]. The average power of a noise source $\overline{e^2(t)}$ can be calculated by using Parseval's theorem. In this paper, the average power is assumed to be stationary

$$\overline{e^2(t)} = \int_0^\infty w(f)\,df$$

$w(f) = $ one-sided spectral density. (1)

By convention in noise analysis, the spectral density $w(f)$ is defined in terms of positive frequency only. The mean square value $\overline{e^2(t)}$ is often expressed by (2) as a function of the noise frequency bandwidth Δf

$$\overline{e^2(t)} = w(f_n)\Delta f. \qquad (2)$$

The bandwidth Δf used in all noise calculations is the bandwidth of an ideal bandpass circuit that has a rectangular response of the same area and peak value as the

Manuscript received May 27, 1976.
The author is with Machine Tool Works, Oerlikon-Bührle Ltd., Zürich, Switzerland.

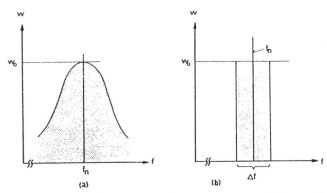

Fig. 2. Definition of noise bandwidth Δf. Shaded area of (a) must be equal to shaded area of (b). For obtaining the area (a) it has to be integrated over the whole frequency range $0 < f < \infty$. (a) Response of circuit. (b) Ideal rectangular response.

sources [21], [22]. These noise sources are given by their spectral and cross spectral densities. Fig. 3 shows different representations of the equivalent external noise sources of a two-port. The correlation of these noise sources can be characterized by a correlation factor. For example, the configuration of Fig. 3(e) has a correlation factor γ given by (3)

$$\gamma = \gamma_1 + j\gamma_2 = \frac{\overline{i_F v_F^*}}{\sqrt{\overline{i_F^2}\ \overline{v_F^2}}} \tag{3}$$

$* =$ complex conjugate;

$\gamma_1, \gamma_2 =$ real and imaginary part of the correlation factor.

γ represents the ratio of the cross power to the geometric mean-value of the two noise powers. It may be assumed that one of the two correlated noise sources consists of a completely correlated component and an uncorrelated component. However, this formal division is arbitrary and can be made in many different ways. The argument of the complex function which relates completely correlated noise sources refers to the *phase difference* of these source. But the phases of the two noise disturbances are not definite.

For many systems, the noise wave parameter representations of Fig. 4 are more suitable [23]. The configurations of Fig. 3 can be transformed to the configurations of Fig. 4. To exemplify such a transformation, the formula for transforming Fig. 3(e) to Fig. 4(e) is derived.

In the case of Fig. 5(a) the following four equations (4)–(7) hold:

$$V_1 = V_2 \tag{4}$$

$$I_1 = I_2 + i_F \tag{5}$$

$$a_2 = a_1 + a_c' \tag{6}$$

$$b_1 = b_2 + b_c'. \tag{7}$$

Equations (8) and (9) are valid if the input line of the two-port shows the characteristic line impedance Z_o

$$a_1 = \frac{1}{2\sqrt{Z_o}}(V_1 + Z_o I_1) \tag{8}$$

$$b_1 = \frac{1}{2\sqrt{Z_o}}(V_1 - Z_o I_1). \tag{9}$$

Similar equations can be given for the output port. As a result, the noise wave sources a_c' and b_c' are obtained as a function of i_F and Z_o if the same line impedance Z_o occurs at the output port

$$a_c' = -\frac{1}{2} i_F \sqrt{Z_o} \tag{10}$$

$$b_c' = -\frac{1}{2} i_F \sqrt{Z_o}. \tag{11}$$

By the same procedure, a_c'' and b_c'' of Fig. 5(b) can be determined. The transformation formulas (12) and (13)

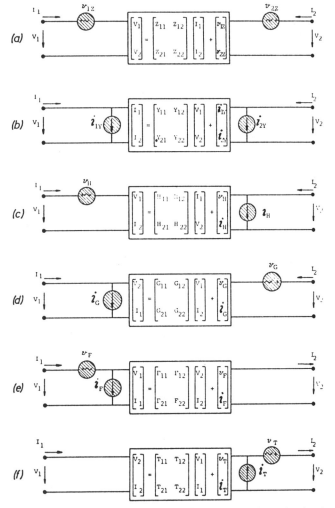

Fig. 3. Different representations of equivalent external correlated noise sources. Parameters G, F, T are also named with C, A, B (see definitions of different countries).

circuit under investigation (Fig. 2). Therefore the noise bandwidth Δf is not the same as the 3-dB bandwidth of a tuned circuit. The internal noise sources of a two-port network can be replaced by two external correlated noise

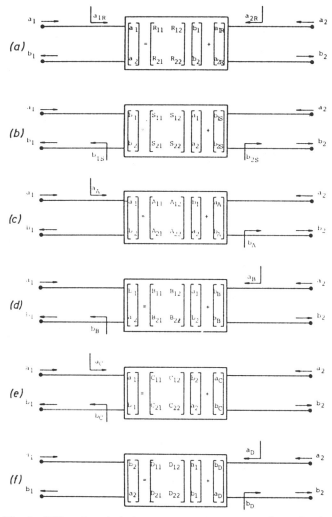

Fig. 4. Different equivalent wave parameter representations of noisy two-port. Parameters A, B, C, D are also named with other letters, e.g., T for C (see definitions of different countries).

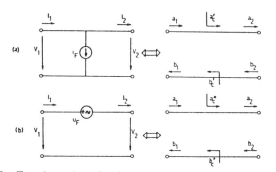

Fig. 5. Transformation of noise source representation to noise wave representation.

are obtained by linear superposition

$$a_c = a_c' + a_c'' = -\frac{1}{2}\left(\frac{v_F}{\sqrt{Z_o}} + i_F \sqrt{Z_o}\right) \quad (12)$$

$$b_c = b_c' + b_c'' = \frac{1}{2}\left(\frac{v_F}{\sqrt{Z_o}} - i_F \sqrt{Z_o}\right). \quad (13)$$

Using the equivalent circuits (Figs. 3 and 4), the noise figure of any linear two-port can be determined. The noise figure definition [8], [15], [16] is given by (14)

$$F = \left.\frac{S_{\text{in}}/N_{\text{in}}}{S_{\text{out}}/N_{\text{out}}}\right|_{T_o = 290 \text{ K}}$$

F = noise figure

T_o = standard reference temperature of the (14)
input termination (290 K)

$S_{\text{in}}, S_{\text{out}}$ = input and output signal powers

$N_{\text{in}}, N_{\text{out}}$ = input and output noise powers.

Since the ratio $S_{\text{in}}/N_{\text{in}}$ is independent of the mismatch at the input and output for S_{in} and N_{in} the corresponding *available input signal* and *noise power* can be introduced in (14). S_{out} and N_{out} represent the *delivered output signal* and *noise power*. The ratio $S_{\text{out}}/N_{\text{out}}$ is independent of the load if the load is noisefree. F is therefore independent of the load. The noise figure with respect to a small bandwidth is called spot noise figure. For most transmission systems the use of the definition of the signal to noise ratio (S/N) is rather straightforward when both terms are expressed in average power of rms values. There are some open questions concerning the correlation between the video signal-to-noise ratio and the RF noise figure because in video signals, (S/N) is defined as the ratio of the peak-to-peak amplitude of the monochrome signal to the rms amplitude of the noise [24].

An equivalent description of the noise performance is given by the input noise temperature. The relation between these two descriptions is shown in (15)

$$T_e = T_o(F - 1) \quad (15)$$

T_o = standard input reference temperature (290 K);

T_e = effective input noise temperature.

Haus and Adler [25] introduced the definition of the exchangeable power P_e in order to include circuits with negative source resistances. P_e is the stationary value (extremum) of the power output from the source, obtained by arbitrary variation of the terminal current (or voltage). Its magnitude represents the maximum power that can be pushed into the terminals by suitable choice of the complex terminal current. Based on the exchangeable power, the exchangeable gain G_e can also be defined.

Using the quantities P_e and G_e, it can be shown that for negative source resistances the noise figure is smaller than unity and for positive source resistances larger than unity. Unfortunately, this exchangeable power concept collapses if the source admittance is reactive for then P_e becomes infinite. Moreover, if the negative source resistance equals the real part of the input impedance, the system is unstable and the signal wave source does not govern the exchange of power [26].

To characterize the composite noise performance of cascaded two-ports (Fig. 6), both the noise figure and the

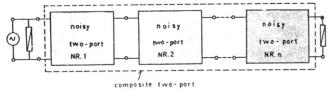

Fig. 6. For noise performance characterization of cascaded two-port eight parameters are used for every two-port, namely, four noise and four gain parameters.

exchangeable gain (available gain for positive real parts of the source admittance) must be known. In this case *more than four noise parameters* are used. In [8], Friis demonstrates the validity of the following equation for positive source resistances:

$$F = F_1 + \frac{F_2 - 1}{G_{a_1}} + \frac{F_3 - 1}{G_{a_1} G_{a_2}} + \cdots \quad (16)$$

first   second   third
stage   stage   stage

F = noise figure of the cascade;

$F_1, F_2, F_3, \cdots$ = noise figure of the two-port numbered from the input to the output of the cascade;

$G_{a_1}, G_{a_2}, \cdots$ = available gain of the different two-ports, numbered as the noise figure;

Equation (16) shows that the noise figure is usually of greater importance in the input stages with gain, although in some cases overload specifications must also be made. It is important to note that the available gain rather than the transducer gain must be used in (16). The available gain G_a of a two-port can be expressed by (17) [27]–[29]

$$\frac{1}{G_a} = \frac{1}{G_{a_{max}}} + \frac{R_g}{\text{Re}(Y_s)}\left((\text{Re}(Y_s) - \text{Re}(Y_{so}))^2 + (\text{Im}(Y_s) - \text{Im}(Y_{so}))^2\right) \quad (17)$$

$G_{a_{max}}$ = maximum available gain;

$\text{Re}(Y_s), \text{Im}(Y_s)$ = real and imaginary part of the source admittance;

$\text{Re}(Y_{so}), \text{Im}(Y_{so})$ = optimal real and optimal imaginary part of the source admittance. These values are obtained if $\text{Re}(Y_s)$ and $\text{Im}(Y_s)$ are varied until $G_a = G_{a_{max}}$.

R_g is defined by (18)

$$R_g = \frac{\text{Re}(Y_{22})}{|Y_{21}|^2} \quad (18)$$

$\text{Re}(Y_{22})$ = real part of the output admittance of the two-port (short-circuited input);

Y_{21} = transfer admittance.

Equation (17) demonstrates how the available gain differs from the maximum available gain if the source admittance is not optimum. The available gain can be characterized by the four gain parameters $G_{a_{max}}$, R_g, $\text{Re}(Y_{so})$, $\text{Im}(Y_{so})$. For negative source resistances, the exchangeable quantities previously discussed can be introduced in (16) and (17).

It follows that the noise performance of a two-port in a cascade (Fig. 6) must be characterized by *eight parameters*, namely, by the four noise parameters and the four gain parameters.

Another noise quantity M, the *noise measure*, is defined in (19)

$$M = \frac{F - 1}{1 - \frac{1}{G_a}} \quad (19)$$

F = noise figure;
G_a = available gain.

The noise measure is useful with the cascade of two-ports for the selection of the optimal amplifier containing various stages. For negative source resistances, the exchangeable noise figure F_e and the exchangeable gain G_e must be introduced in (19) instead of F and G_a. Haus and Adler [25] prove that the "best" cascade order of two amplifiers with respect to the noise performance is the one with the amplifier characterized by the lower noise measure at the input.

As stated in [25], this is only true under the condition that the available gain and the noise figure of each amplifier do not change when the order of cascading is reversed.

This assumption is not always satisfied since F and G_a are dependent on the signal source impedance at the network input. The following procedure is therefore proposed. If a cascade consisting of two two-ports is considered, (20) is valid if the amplifier x precedes the amplifier y, and (21) is valid for the reverse order

$$F_1(Y_s) = F_x(Y_s) + \frac{F_y(Y_s') - 1}{G_{ax}(Y_s)} \quad (20)$$

$$F_2(Y_s) = F_y(Y_s) + \frac{F_x(Y_s') - 1}{G_{ay}(Y_s)} \quad (21)$$

Y_s = source admittance at the cascade input;
Y_s' = output admittance of the first stage;
F_1 = noise figure of the cascade for the two-port order (x before y)
F_2 = noise figure of the cascade for the two-port order (y before x).

The minimization of the noise figures F_1 and F_2 can be executed numerically. Visual insight can be gained by using design graphs as shown in [27], [28], [29].

The optimal cascade order for two amplifiers characterized by low gain is not necessarily obtained if the

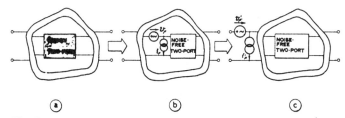

Fig. 7. Transformation of equivalent noise sources v_F and i_F to v'_F and i'_F of the configuration (c). Surrounding network is supposed to be noisefree.

source admittance of the cascade is chosen corresponding to the minimal noise figure of one amplifier or if the source admittance is changed in order to achieve the maximum available gain. To obtain the minimal noise figure, the source admittance must correspond to a tuning that lies between lowest noise and maximum power tuning. In addition, the correct cascade order must be selected. If F_1 and F_2 are minimized separately this order can be determined by (20) and (21). This procedure is valid under all conditions. The noise measure M will serve only as an indicator of the extent to which a given design fails to achieve its best noise performance. Since there are signal representations in which neither the noise measure nor the exchangeable noise power can be interpreted in elementary physical quantities relevant for the system, Bosma concludes in his study [26] that the characteristic noise temperature (see eq. (15)) is the absolute, independent, and unique measure of the spot-noise intensity for all kinds of linear systems. This conclusion can only be accepted with regard to the shown physical interpretation of the noise behavior. In connection with the said physical interpretation, it is interesting to note that a new sign convention of the noise temperature is discussed in [26] by introducing the negative sign for the noise temperature of amplifiers. However the proposed sign convention is not standardized. Nevertheless the noise figure corresponding to the IEEE standard is an equivalent number for noise calculations. It should be emphasized that the noise figure (or the noise temperature) alone is not sufficient for the noise characterization of a device. The circuit and device designer often encounters the problem of determining the noise performance of a more or less complex circuit, consisting of different subdevices, each with a known noise performance. These complex circuits can be more complicated than a cascade. In this case the transformation of the noise parameters by imbedding two-ports can be obtained either by combining the different two-ports with series–series, parallel–parallel, series–parallel, and parallel–series connections or the noise parameters can be calculated directly by using more efficient procedures. The first method enhances the physical understanding. The second method to be described, is desirable for very large circuits if the calculation time has to be considered.

In Fig. 7, the transformation of the noise parameters by a noise-free imbedding network is represented. The noisy two-port is transformed to the configuration of Fig. 3(e).

Applying the noise figure formula (14) the following (22) can be derived for this configuration of Fig. 3(e)

$$F = 1 + \frac{G_n}{\text{Re}(Y_s)} + \frac{R_f}{\text{Re}(Y_s)} |Y_s + Y_{\text{cor}}|^2 \qquad (22)$$

G_n = equivalent noise conductance;

Y_s = source admittance.

R_f represents the equivalent noise resistance. Using the Nyquist formula (23), the mean square value of v_F can be given as a function of this resistance R_f [30], [31]

$$\overline{v_F^2} = 4kTR_f \Delta f. \qquad (23)$$

The noise current source i_F can be separated into an uncorrelated part $(i_F)_{\text{un}}$ and a correlated part $v_F Y_{\text{cor}}$

$$i_F = (i_F)_{\text{un}} + v_F Y_{\text{cor}}. \qquad (24)$$

The mean square current value of the uncorrelated part can be represented as a function of the equivalent noise conductance G_n

$$\overline{(i_F)_{\text{un}}^2} = 4kTG_n \Delta f. \qquad (25)$$

The correlation admittance Y_{cor} is expressed by (26)

$$Y_{\text{cor}} = \frac{\overline{i_F v_F^*}}{\overline{v_F^2}}. \qquad (26)$$

The defined four parameters G_n, R_f, Re (Y_{cor}), and Im (Y_{cor}) of the inner two-port of Fig. 7 can be transformed to those of the composite network by using the four complex small signal parameters of this inner two-port as well as the small signal parameters of the surrounding network [22]. Therefore it can be concluded that the noise performance of the inner two-port generally surrounded by a network has to be characterized by 12 parameters, namely by its four noise parameters (as for example Y_{cor}, R_f, G_n) and its four complex two-port parameters.

There does not exist any relationship among these 12 parameters. The source admittance should not be ignored even if for simpler investigations the noise is only described by independent noise sources [32]. The equations derived in [22] are only valid if the scattering matrix of the original inner noisy two-port (Fig. 6) is nonsingular. The physical interpretation of this condition is given by the fact that the total system formed by the original and the imbedding network must not have undamped internal resonances [26].

For the different two-port configurations, the transformation of the correlation functions can also be given in matrix form. It should be mentioned that there are other different correlation function definitions instead of Y_{cor} [33], but all of these possible representations lead to the same end result.

It is often emphasized that feedback does not affect noise figure or the noise measure. This is true only under certain conditions in which such important factors as the noise originating in the feedback loop are neglected [34]–[41]. It is advisable to use the exact transformation equations and the transformation formulas for the two-port parameters if there is any doubt.

Finally, it should be noted that an amplifier designer must consider not only the noise performance but also factors such as distortion, stability, temperature dependence, phase characteristic, transient response, etc. It is possible to include all of these factors in a weighted object function. This function can be optimized with respect to the two-port parameters using computer-aided methods [42], [43].

III. Noise Calculations

In the frequency domain, the procedure using the adjoint network concept [44], [45] can be applied not only for *uncorrelated* or *fully correlated* noise sources but also for *partially correlated* noise sources. The determination of the noise figure of a network is derived in a more general way than shown in [46] with the special transistor example. For a stable system, a system function $H_n(f)$ can be defined

$$H_n(f) \triangleq \frac{A_n^o(f)}{A_n^i(f)} \qquad (27)$$

where $A_n^i(f)$ is the complex amplitude of the nth-sinusoidal input signal and $A_n^o(f)$ is the resulting amplitude at the output port. Every noise source is thought to be at an input of an n-port circuit. The output noise power spectral density $S_o(f)$ is obtained with (28) [20]

$$S_o(f) = \sum_{n=1}^{N} \sum_{m=1}^{N} H_n(f) H_m^*(f) S_{nm}(f) \qquad (28)$$

$* =$ conjugate complex value;

$S_{nm} =$ spectral densities for $m = n$ and cross spectral densities for $m \neq n$ of the different noise sources;

$N =$ number of input ports.

In a stable fixed-parameter linear system, the system function $H_n(f)$ can be expressed in terms of the various short-circuit transfer admittances (eq. (29)) if only noise voltage sources are considered at the different input ports as well as at the output ports. In this case $A_n^o(f)$ and $A_n^i(f)$ (eq. (27)) represent voltage amplitudes

$$H_n(f) = -\frac{Y_{on}(f)}{Y_{oo}(f)} \qquad \text{(network with noise voltage sources)} \qquad (29)$$

$Y_{on}(f) =$ ratio of the amplitude of the short-circuit current flowing in the output terminal pair to the complex amplitude of the voltage applied to the nth port when all other terminal pairs are short-circuited;

$Y_{oo}(f) =$ short-circuit driving-point admittance of the output terminal pair (all other ports are short-circuited).

Similarly $H_n(f)$ is defined for networks with noise current sources only using the open circuit transfer impedances (see eq. (30))

$$H_n(f) = -\frac{Z_{on}(f)}{Z_{oo}(f)} \qquad \text{(network with noise current sources)} \qquad (30)$$

$Z_{on}(f) =$ ratio of the complex amplitude of the open-circuit voltage in the output port to the complex amplitude of current applied to the nth port when all other terminal pairs are open-circuited;

$Z_{oo}(f) =$ open-circuit driving-point impedance of the output terminal pair (all other ports are open circuited).

The output spectral density $S_o(f)$ can be expressed by (31) for networks with noise voltage sources (see [20])

$$S_o(f) = \sum_{n=1}^{N} \sum_{m=1}^{N} \frac{Y_{on} Y_{om}^*}{|Y_{oo}|^2} S_{nm}(f) \qquad (31)$$

and by (32) for networks with noise current sources

$$S_o(f) = \sum_{n=1}^{N} \sum_{m=1}^{N} \frac{Z_{on} Z_{om}^*}{|Z_{oo}|^2} S_{nm}(f). \qquad (32)$$

Using these definitions the spot noise figure F, which is only defined with respect to two-ports, can be expressed by (33)

$$F = 1 + \frac{S_o''(f)}{S_o'(f)} \qquad (33)$$

$S_o''(f) =$ output noise power spectral density calculated with all internal noise sources (without the noise generator contribution at the input port);

$S_o'(f) =$ output noise power spectral density (with only the noise generator at the input port).

The interreciprocity concept can be used when applying (31), (32), (33). The application of the adjoint network concept can be illustrated by using Fig. 8.

Considering only the port variables between the original and the adjoint network, the following relation (eq. (34)) can be derived [45] with respect to Fig. 8:

$$\sum_{p=1}^{N_p} (VI' - V'I) = 0 \qquad (34)$$

$N_p =$ total number of ports.

This relation is valid if a sign convention introduced in [46] is applied in the original and in the adjoint network. If $V = V'$ in Fig. 8(a) it follows that $I = I'$ and if $I = I'$ in Fig. 8(b) it follows that $V = V'$. The transadmittances (or transimpedances) with respect to the noise source ports as well as the input driving point admittance (or input driving-point impedance) can thus be obtained by one single network analysis of the adjoint network if the network contains only voltage sources (or current sources), since

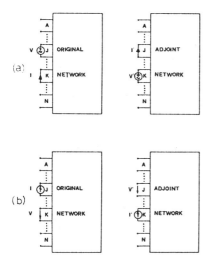

Fig. 8. (a) Original and adjoint network with voltage excitation. (b) Original and adjoint network with current excitation.

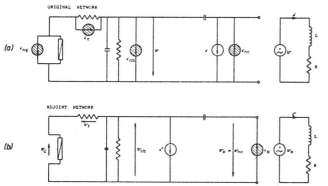

Fig. 9. (a) Simplified transistor equivalent circuit containing noise sources. (b) Adjoint network of (a) with unique current source at output port.

the admittances and impedances are determined by dividing the corresponding voltage and current values. This result can be illustrated with a practical example using the simplified equivalent circuit (Fig. 9) of a transistor containing only noise current sources [19], [47]–[50]. The voltages at the different ports of the adjoint network are calculated with respect to the current source i_N at the output port (Fig. 9(b)). In the adjoint network, the noise sources are omitted and only the excitation at the output is considered. The voltages in the adjoint network are calculated only at the ports determined by the corresponding noise sources in the original network. Thus the different transfer and input driving-point impedances are obtained by one single network analysis. As a consequence of (31)–(33) the noise figure F is given by

$$F(f) = 1 + \frac{\frac{|Z_{o1}|^2}{|Z_{oo}|^2} S_{11}(f) + \frac{|Z_{o2}|^2}{|Z_{oo}|^2} S_{22}(f) + 2\,\mathrm{Re}\left(S_{12}(f)\frac{Z_{o1}Z_{o2}^*}{|Z_{oo}|^2}\right) + \frac{|Z_{o3}|^2}{|Z_{oo}|^2} S_{33}(f)}{\frac{|Z_{o4}|^2}{|Z_{oo}|^2} S_{44}(f)} \quad (35)$$

$$S_{11}(f) = 4kT\,\mathrm{Re}(Y_{11e'}) + 2e\left[I_c - \frac{I_E}{m_E}\right] \quad (36)$$

$$S_{22}(f) = 2eI_c \quad (37)$$

$$S_{12}(f) = 2eI_c - 2kTY_{21e'} \quad (38)$$

$$S_{33}(f) = 4kT\frac{1}{R_{\mathrm{base}}} \quad (39)$$

$$S_{44}(f) = 4kT\,\mathrm{Re}(Y_s) \quad (40)$$

$\mathrm{Re}(Y_{11e'})=$ real part of the intrinsic base–emitter input admittance;
$Y_{21e'}=$ transadmittance;
$e=$ absolute charge of the electron;
$I_c=$ collector dc current;
$I_E=$ emitter dc current;
$m_E=$ high injection factor;
$*=$ conjugate complex.

These given spectral densities are "one-sided." S_{11} represents the spectral density of the base noise source $\overline{i_{nb}^2}$, S_{22} that of the collector noise source $\overline{i_{nc}^2}$, S_{33} that of the base resistance noise source $\overline{i_r^2}$ and S_{44} that of the noise generator $\overline{i_{ng}^2}$. S_{12} considers the correlation of $\overline{i_{nb}^* i_{nc}}$. The impedances $Z_{o1}, Z_{o2}, Z_{o3}, Z_{o4}$ represent the transimpedances with respect to the port of the base noise source, the collector noise source, the base resistance noise source, and the noise generator. The validity of the equation $S_{12}(f) = S_{21}^*(f)$ is demonstrated in [20]. Moreover, the sum of two conjugate complex numbers can be expressed by the addition of their equal real parts.

For microwave applications, the noise wave parameter representation is usually the most suitable. Using the scattering parameter configuration of Fig. (2b), the transformation of the correlated noise waves b_{1s}' and b_{2s} is also obtained by separating these waves into an uncorrelated and a correlated part. Then the actual theories [42], [51]–[53] can be used in an efficient computer analysis developed especially for microwaves.

IV. Noise Parameter Measurement

The experimental determination of the noise figure as a function of the frequency is often based on the fact that the noise figure of a device can also be interpreted as the ratio of the total noise power available at the output port when the input termination is at 290 K to that portion of the total available output power engendered by the input termination and delivered to the output via the primary signal channel. The definition of the noise factor is not restricted to the condition that the equipment whose noise

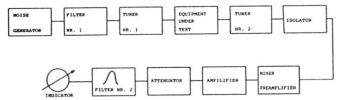

Fig. 10. Equipment arrangement to measure noise figure.

figure is to be measured be linear. But the measurement technique requires a linear measured device. Therefore the input signal from the noise generator has to be carefully limited to avoid nonlinearities. The following (41) is often suitable for measurements [54]–[56]. It is obtained by using (22)

$$F = F_{min} + \frac{R_n}{\text{Re}(Y_s)} \{ (\text{Re}(Y_s) - \text{Re}(Y_{so}))^2$$
$$+ (\text{Im}(Y_s) - \text{Im}(Y_{so}))^2 \} \quad (41)$$

F_{min} = minimal noise figure;

R_n = parameter with units of resistance, represents a measure for the increase of the noise figure, if the source admittance differs from the optimal one;

Y_s = source admittance;

Y_{so} = optimal source admittance; this admittance is necessary for obtaining the best noise figure F_{min};

Re, Im = real and imaginary part.

Equation (41) can be brought into the form of (42)

$$F_i = a + \frac{1}{\text{Re}(Y_{si})} b + \frac{|Y_{si}|^2}{\text{Re}(Y_{si})} c - 2 \frac{\text{Re}(Y_{si})}{\text{Im}(Y_{si})} d \quad (42)$$

$$a = F_{min} - 2R_n \text{Re}(Y_{so}) \quad (43)$$

$$b = R_n (\text{Re}^2(Y_{so}) + \text{Im}^2(Y_{so})) \quad (44)$$

$$c = R_n \quad (45)$$

$$d = R_n \text{Im}(Y_{so}). \quad (46)$$

The four new parameters a, b, c, d are linearly related to the noise figure. If four measurements are made with four different source admittances Y_{si}, the four quantities a, b, c, d are determined and F_{min}, R_n, $\text{Re}(Y_{so})$, $\text{Im}(Y_{so})$ can be calculated. For higher measurement accuracy more than four points should be used. In [54] a similar method is demonstrated while keeping the real part of the source admittance constant.

To illustrate some techniques used, a possible equipment arrangement to measure the noise figure is shown in Fig. 10 [57]. The noise generator in front of the equipment can be a hot–cold, semiconductor, or a gas-discharge noise source. Its precise calibration as a function of the frequency is essential for selecting the best generator. The second stage of Fig. 10 represents a filter. It suppresses the unwanted frequency components. The tuner Nr. 1 allows adjustment of the source admittance in order to measure the noise figure as a function of the source admittance corresponding to (41) and (42). Tuner Nr. 2 serves as a matching device. The source admittance of the following stages is held constant by using an isolator or an impedance transformer (cathode follower) since the noise figure of the measurement equipment depends on the output admittance of the test device. The following mixer-preamplifier stage is necessary for very high frequencies and can be omitted in the lower frequency ranges. The attenuator serves as a noise power regulator in connection with the 3-dB measurement method. An accurate determination of the spectral noise figure distribution implies a bandwidth of the bandpass filter as small as possible compared with the corresponding frequency. The noise indicator at the end of Fig. 10 must be able to indicate quadratic values. It is necessary that the whole measurment equipment introduce negligible noise. To reduce the noise figure of the measurement equipment, it may be necessary to put a low-noise parametric amplifier in front of the mixer stage. Using the mixer stage requires that the image frequency signals be suppressed or taken in account by calculation [58]. In [59] a summary of measurement techniques is shown (tangential sensitivity method, three-decibel method, automatic noise factor meter, CW-method, Y-factor method).

In the lower frequency ranges, the time constant of the indicator instrument must be enlarged to avoid oscillation at lower frequencies since the filter bandwidth has to be small compared to the frequency under test. For a mean error of one percent, the following (47) is approximately valid [60], [61]

$$T\Delta f = 10^3 \to 10^4. \quad (47)$$

Since the necessary time constant T can be very large, other procedures have been developed [62], [63]. Accurate methods are based on the measurement of the autocorrelation function using correlator instruments.

The possibility of determining the noise figure at negative source admittances or negative input and output admittances is also essential for the designer [64], [65]. Therefore three cases are distinguished, namely, a) the device under test has positive real parts of the input and output admittance, b) the real part of the input admittance is negative, and c) the real part of the output admittance is negative.

Oscillators have been treated in [66], [67]. Therefore the condition $\text{Re}(Y_s) \neq \text{Re}(Y_{in})$ has to be considered in the following description. Also the case $\text{Re}(Y_s) = 0$ is excluded.

Procedure for Case a)

The noise figure with respect to negative source conductances is obtained by measuring at positive conductances and by calculating the exchangeable noise figure at negative source conductances using (41).

Procedure for Case b)

The noise figure is measured at positive source conductances containing the absolute value of the source conductance larger than the absolute value of the input conductance. Then F is calculated using (41).

Procedure for Case c)

If the output conductance of the test device is negative, a further two-port with positive input and output conductances can be connected at the output of the first stage. The noise figure F_e of the total configuration is then measured at positive source conductances. After the determination of the corresponding negative output admittance of the first two-port which represents the source admittance of the second two-port, the second stage is measured itself at positive source conductances. The exchangeable noise figure of the second stage F_{e2} is then obtained by calculation for negative source conductances. Finally, (48) is used for calculating the noise figure F_{e1} of the test device

$$F_{e1} = F_e - \frac{F_{e2}-1}{G_{e1}}. \quad (48)$$

G_{e1} represents the exchangeable gain of the first stage. It can be determined by measuring the transducer gain G_{T1} and the output and load admittances Y_{out} and Y_L of this stage. Using (49) G_{e1} can be calculated

$$G_{e1} = G_{T1} \frac{|Y_{out} + Y_L|^2}{4 \operatorname{Re}(Y_{out}) \operatorname{Re}(Y_L)}. \quad (49)$$

Conclusion

The noise performance of linear two-ports can be characterized generally by 12 parameters, namely by four noise and four complex two-port parameters. The data sheets of the different components should include all of these quantities. The different parameters should be measured accurately using the methods described or they must be at least theoretically well known. Knowing the exact parameter values, the optimal design can be found rapidly by using the methods discussed. Finally it must be stated that the noise figure concept is not always sufficient. For more difficult problems, the statistical functions have to be applied. In addition, phase fluctuations should be included since only amplitude fluctuations have been discussed in this paper. It would be interesting to discuss the theory under this aspect in a further contribution.

Acknowledgment

The author wishes to acknowledge the IEEE reviewers whose comments and suggestions have led to significant improvements in this article.

References

[1] R. Brown, "Mikroskopische Beobachtungen über die in Pollen von Pflanzen enthaltenen Partikeln, und über das allgemeine Vorkommen aktiver Moleküle in organischen und anorganischen Körpern," *Ann. Phys.*, vol. 14, p. 294, 1828.
[2] A. Einstein, "Zur Theorie der Brownschen Bewegung," *Ann. Phys.*, vol. 19, pp. 371–381. 1906.
[3] M. von Smoluchovski, "Zur Kinetischen Theorie der Brownschen Molekularbewegung und der Suspensionen," *Ann. Phys.*, vol. 21, pp. 756–780, 1906.
[4] J. Perrin, "L'agitation moléculaire et le mouvement brownien," *Comptes rendues Acad. Sci. Paris*, vol. 146, p. 967, 1908.
[5] G. L. de Haas-Lorentz, *Die Brownsche Bewegung*, Vieweg, Braunschweig, 1913.
[6] R. E. Burgess, "Noise in receiving aerial system," *Proc. Phys. Soc.*, vol. 53, 293–304, 1941.
[7] D. O. North, Discussion of "Noise figures of radio receivers" by H. T. Friis, *Proc. IRE*, vol. 33, 125–127, 1945.
[8] H. T. Friis, "Noise figure of radio receivers," Proc. IRE, vol. 32, pp. 419–422, 1944.
[9] H. Goldberg, "Some notes on noise figures," *Proc. IRE*, vol. 36, p. 1205, 1948.
[10] H. A. Haus and R. B. Adler, "Optimum noise performance of linear amplifiers," *Proc. IRE*, vol. 46, pp. 1517–1533, 1957.
[11] I. A. Harris, "Errors and uncertainties in microwave measurements," *Proc. IEE*, supplement no. 23, vol. 109B, p. 841, 1962.
[12] E. P. McCurley and C. Blake, "A simple approximate expression for converting directly from noise figures in dB to noise temperature," *Microwave Journal*, vol. 4, p. 79, 1961.
[13] K. A. Norton, Efficient use of the radio spectrum, NBS Tech. Note 158, 1962.
[14] Standards on Electron Devices, "Methods of Measuring Noise," *Proc. IRE*, vol. 41, pp. 890–896, 1953.
[15] IRE Standards "Methods of measuring noise in linear two-ports," *Proc. IRE*, p. 68, 1960.
[16] IRE Subcommittee 7-9 on Noise, "Description of the noise performance of amplifiers and receiving systems," *Proc. IEEE*, vol. 51, pp. 436–422, 1963.
[17] M. J. O. Strutt, *Semiconductor Devices*. New York; Academic Press, 1966.
[18] A. van der Ziel, Noise in solid-state devices and lasers, *Proc. IEEE*, vol. 58, pp. 1178–1206, 1970.
[19] W. R. Bennet, *Electrical Noise*. New York: McGraw Hill, 1960.
[20] W. B. Davenport and W. L. Root, *Random Signals and Noise*. New York: McGraw Hill, 1958.
[21] H. Rothe and W. Dahlke, "Theory of noisy fourpoles," *Proc. IRE*, pp. 811–818, June 1956.
[22] K. Hartmann and M. J. O. Strutt, "Changes of the four noise parameters due to general changes of linear two-port circuits," *IEEE Trans. Electron Devices*, pp. 874–877, Oct. 1973.
[23] K. Hirano, R. Koiwai, and K. Yamamoto, "Matrix representation of linear noisy networks and its transformation," Memoirs of the Faculty of Engineering, Kobe University, Kobe, Japan, No. 14 pp. 63–76, 1967.
[24] J. Simonson, "Analyze TV-noise performance by correlating video signal-to-noise ratio with rf noise figure," *Electronic Design*, pp. 82–87, Sept. 1974.
[25] H. A. Haus and R. B. Adler, *Circuit Theory of Linear Noisy Networks*. New York: Wiley, 1959.
[26] H. Bosma, "On the theory of linear noisy systems, "thesis, Technical University Eindhoven, Eindhoven, The Netherlands, 1967.
[27] H. Fukui, Available power gain, noise figure and noise measure of two-ports and their graphical representations, *IEEE Trans. Circuit Theory*, vol. 13, 137–142, June 1966.
[28] W. Baechtold and M. J. O. Strutt, "Noise in microwave transistors," *IEEE Trans. Microwave Theory and Techniques*, pp. 578–585, Sept. 1968.
[29] V. Zalud, "A chart for solving two-port noise, gain and stability problems," *Archiv der Elektrischen Uebertragung*, A.e.Ue., Vol. 25, pp. 537–539, 1971.
[30] H. Nyquist, "Thermal agitation of electric charge in conductors," *Phys. Rev.*, vol. 32, pp. 110–113, July 1928.
[31] J. G. Linvill and J. F. Gibbons, *Transistors and Active Circuits*. New York: McGraw Hill, 1961.
[32] A. M. Darbin, "Reduce noise in feedback circuits," *Electronic Design*, vol. 25, pp. 72–75, Dec. 1972.
[33] H. Hillbrand and P. Russer, "An efficient method for computer aided noise analysis" *IEEE Trans. Circuits and Systems*, vol. CAS-23, pp. 235–238, Apr. 1976.
[34] W. A. Rheinfelder, *Design of Low-Noise Transistor Input Circuits*. London, England: Iliffe Books, 1964.
[35] A. van der Ziel, *Noise*. New York: Prentice-Hall, 1954.
[36] J. Engberg, "Simultaneous input power match and noise optimization using feedback," *Proc., 4th European Microwave Conf.*, Montreux, Switzerland, Sept. 10–13, 1974.

[37] K. Hartmann, "Small signal and noise behavior of microwave bipolar transistors," presented at the 3rd European Microwave Conf., Brussels, Belgium, Sept. 4–7, 1973.
[38] G. Guekos and M. J. O. Strutt, "Reduction of the excess noise of GaAs diode lasers by optoelectrical feedback," *Proc. IEEE*, pp. 949–950, June 1970.
[39] A. Anastassiou and M. J. O. Strutt, "Effect of source lead inductance on the noise figure of a GaAs FET," *Proc. IEEE*, pp. 406–408, Mar. 1974.
[40] H. Beneking, "Praxis des elektronischen Rauschens," "Bibliographisches Institut, Hochschulskripten Nr. 734, Mannheim, Germany.
[41] S. Iversen, "The effect of feedback on noise figure," *Proc. IEEE*, pp. 540–542, Mar. 1975.
[42] J. W. Bandler and R. E. Seviora, "Wave sensitivities of networks," *IEEE Trans. Microwave Theory and Techniques*, vol. 20, pp. 138–147, 1972.
[43] K. Hartmann and M. J. O. Strutt, "Computer simulation of small-signal and noise behavior of microwave bipolar transistors up to 12 GHz," *IEEE Trans. Microwave Theory and Techniques*, pp. 178–183, Mar. 1974.
[44] S. W. Director and R. A. Rohrer, "Inter-reciprocity and its implications," *Proc. Int. Symp. Network Theory*, Belgrade, Yugoslavia, pp. 11–30, Sept. 1968.
[45] R. Rohrer, L. Nagel, R. Meyer, and H. Weber, "Computationally efficient electronic-circuit noise calculations," *IEEE Journal Solid-State Circuits*, pp. 204–213, Aug. 1971.
[46] K. Hartmann, W. Kotyczka, and M. J. O. Strutt, "Computerized determination of electrical network noise due to correlated and uncorrelated noise sources," *IEEE Trans. Circuit Theory*, vol. 20, pp. 321–322, May 1973.
[47] W. Guggenbühl and M. J. O. Strutt, "Theory and experiments on shot noise in semiconductors junction diodes and Transistors," *Proc. IRE*, June 1957.
[48] B. Schneider and M.J.O. Strutt, "Theory and experiments on shot noise in silicon p–n junction diodes and Transistors," *Proc. IRE*, Apr. 1959.
[49] W. Thommen and M. J. O. Strutt, "Noise figure of UHF-transistors," *IEEE Trans. Electron Devices*, pp. 499–500, Sept. 1965.
[50] M. Schwartz, *Information, Transmission, Modulation and Noise*. New York: McGraw Hill, 1959.
[51] V. A. Monaco and P. Tiberio, "Automatic scattering matrix computation of microwave circuits," *Alta Frequenza*, vol. 39, pp. 165–170, 1970.
[52] F. Bonfatti, V. A. Monaco, and P. Tiberio, "Microwave circuit analysis by sparse matrix techniques," *IEEE Trans. Microwave Theory and Techniques*, vol. 22, pp. 264–269, 1974.
[53] C. Rauscher, "A fast evaluation of s-parameter sensitivities," *Archiv der Elektrischen Uebertragung*, A. E. Ue., vol. 28, pp. 113–114, 1974.
[54] W. Kotyczka, A. Leupp, and M.J.O. Strutt, Computer-aided determination of two-port noise parameters (CADON), *Proc. IEEE*, pp. 1850–1851, Nov. 1970.
[55] J. Lange, "Noise characterization of linear two-ports in terms of invariant parameters," *IEEE Journal Solid-State Circuits*, vol. 2, pp. 37–44, June 1967.
[56] R. Q. Lane, "The determination of device noise parameters," *Proc. IEEE*, pp. 1461–1462, Oct. 1962.
[57] K. Hartmann, "Computer simulation of linear noisy two-port circuits with special consideration of the bipolar transistor up to 12 GHz," Thesis Nr. 9118, Swiss Federal Institute of Technology, Switzerland, 1973.
[58] R. L. Sleven, "Clarification of subtle noise figure measurement problems," AIL Technical Memorandum Nr. 44, Airborne Instruments Laboratory, A Division of Cutler-Hammer Inc., New York.
[59] E. G. Hildner, "A note on noise factor measurement," ESSA Tech. Rep. ERL 133-WPL, U.S. Government Printing Office, Washington, D.C., Oct. 9, 1969.
[60] H. Bittel, "Zur Kennzeichnung von Geräuschen und Rauschspannungen," *Zeitschrift für angew. Physik*, Bd. 4, pp. 137–146, 1952.
[61] S. O. Rice, Mathematical analysis of random noise, *Bell Syst. Tech. J.*, vol. 24, 1945, pp. 46–108.
[62] B. V. Rollin and I. M. Templeton, "Noise in Ge-filaments at very low frequencies," *Proc. Phys. Soc.*, vol. 67B, pp. 271–272, 1954.
[63] U. J. Strasilla and M. J. O. Strutt, "Measurement of white and 1/f noise within burst noise," *Proc. IEEE*, vol. 62, pp. 1711–1713, Dec. 1974.
[64] H. A. Haus, "Noise figure for negative source resistance," *Proc. IEEE*, pp. 2135–2136, Oct. 1962.
[65] A. van der Ziel, "On the noise figure of negative conductance amplifiers," *Proc. IEEE*, pp. 1961–1962, Oct. 1962.
[66] M. T. Vlaardingerbroek, "Theory of oscillator noise," *Electron. Lett.*, vol. 7, pp. 648–650, 1971.
[67] K. Kurokawa, "Noise in synchronized oscillators," *IEEE Trans. Microwave Theory and Techniques*, vol. 16, pp. 234–240, 1968.

Available Power Gain, Noise Figure, and Noise Measure of Two-Ports and Their Graphical Representations

H. FUKUI

Abstract—The expressions for available power gain and noise measure of linear two-ports are introduced in terms of the two-port parameters and the gain and the noise parameters, respectively. Their graphical representations on the source admittance plane with rectangular coordinates are also shown. Furthermore, it is shown that the behavior of available power gain, noise figure, and noise measure can be represented on the Smith-chart or the complex reflection coefficient plane of the source admittance. It is more convenient to investigate the gain and noise performance of amplifiers over a wide range of source admittance in this representation than with rectangular coordinates. As an example of the graphical representation the gain and noise performance of a microwave transistor is illustrated on the Smith-chart.

I. INTRODUCTION

IT IS WELL KNOWN that the noise figure of linear twoports is a function of the source admittance [1]. The noise figure of a linear two-port driven by a signal source with an admittance, $Y_s = G_s + jB_s$, can be expressed as

$$F = F_{\min} + \frac{R_{ef}}{G_s}\{(G_s - G_{of})^2 + (B_s - B_{of})^2\} \quad (1)$$

where $F_{\min}$ is the minimum noise figure, R_{ef} is a parameter with units of resistance, and G_{of} and B_{of} are the particular source conductance and susceptance, respectively, which produce $F_{\min}$. Thus, the noise figure characteristics of linear two-ports can be completely described by the four noise parameters, $F_{\min}$, R_{ef}, G_{of}, and B_{of}. Furthermore, a graphical representation of the noise figure on a rectangular coordinate system of source admittance has been given by Rothe and Dahlke [2].

In recent years, the noise measure has been proposed by Haus and Adler [3] as a more significant parameter than the noise figure when considering cascaded two-ports. The noise measure M is defined as [4]

$$M = \frac{F - 1}{1 - \dfrac{1}{G_a}} \quad (2)$$

where G_a is the available power gain of the two-port.

Since G_a and F are functions of source admittance Y_s, M will also be a function of Y_s. However, this situation has not previously been made clear. This paper will consider the problem and show that constant-M loci can be mapped in the source admittance plane. Another purpose of this paper is to demonstrate how the constant G_a, F, and M loci appear on the Smith-chart representation.

II. AVAILABLE POWER GAIN

A. The Available Power Gain of Two-ports

The available power gain G_a of the two-port shown in Fig. 1 is defined as the ratio of the available power at

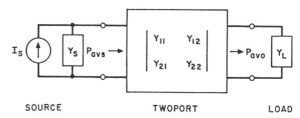

Fig. 1. Available power gain of a twoport.

the output P_{avo} to the available power from the source P_{avs}, i.e.,

$$G_a = \frac{P_{avo}}{P_{avs}}. \quad (3)$$

G_a is a function of the two-port parameters, such as y-parameters, and the source admittance Y_s.

Using the same procedure given by Linvill and Gibbons [5], G_a is expressed as follows.

$$G_a = \frac{|y_{21}|^2 G_s}{g_{22}|y_{11} + Y_s|^2 - \text{Re}\,[y_{12}y_{21}(y_{11} + Y_s)^*]}. \quad (4)$$

B. Alternate Expression for Available Power Gain

From the expression for G_a its maximum value, $G_{a_{\max}}$, is derived for the particular source admittance $Y_{og} = G_{og} + jB_{og}$ as follows.

$$G_{a_{\max}} = \left|\frac{y_{21}}{y_{12}}\right| \frac{1}{k + \sqrt{k^2 - 1}} \quad (5)$$

$$G_{og} = \frac{|y_{12}y_{21}|}{2g_{22}}\sqrt{k^2 - 1} \quad (6)$$

$$B_{og} = -b_{11} + \frac{\text{Im}\,(y_{12}y_{21})}{2g_{22}} \quad (7)$$

where

$$k = \frac{2g_{11}g_{22} - \text{Re}\,(y_{12}y_{21})}{|y_{12}y_{21}|}. \quad (8)$$

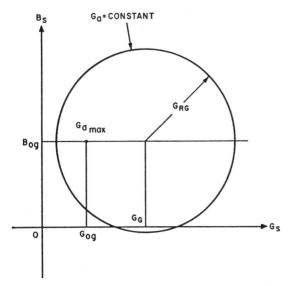

Fig. 2. Constant available power gain locus on rectangular source admittance plane.

Now (4) is rewritten using a set of gain parameters, $G_{a\,max}$, G_{og}, B_{og}, and a parameter R_{eg}:

$$\frac{1}{G_a} = \frac{1}{G_{a\,max}} + \frac{R_{eg}}{G_s}[(G_s - G_{og})^2 + (B_s - B_{og})^2] \quad (9)$$

where

$$R_{eg} = \frac{g_{22}}{|y_{21}|^2}. \quad (10)$$

It should be noticed that (9) is quite similar to the expression for noise figure given in (1). This suggests that a graphical representation similar to that for the noise figure can be expected for the available power gain.

C. A Graphical Representation of Available Power Gain on the Rectangular Source Admittance Plane

Rearranging (9), the following equation can be derived:

$$(G_s - G_G)^2 + (B_s - B_G)^2 = G_{RG}^2 \quad (11)$$

where

$$G_G = G_{og} + \frac{1}{2R_{eg}}\left(\frac{1}{G_a} - \frac{1}{G_{a\,max}}\right) \quad (12)$$

$$B_G = B_{og} \quad (13)$$

$$G_{RG} = \left[\frac{G_{og}}{R_{eg}}\left(\frac{1}{G_a} - \frac{1}{G_{a\,max}}\right) + \frac{1}{4R_{eg}^2}\left(\frac{1}{G_a} - \frac{1}{G_{a\,max}}\right)^2\right]^{1/2}. \quad (14)$$

Equation (11) represents a family of circles on the rectangular coordinates with G_s as the abscissa and B_s as the ordinate, as shown in Fig. 2. Each circle represents a constant G_a locus and has its center at the point (G_G, B_G) and a radius of G_{RG}. G_G and G_{RG} depend upon the value of G_a, but B_G is independent of G_a. Therefore, the locus of the center points for constant G_a becomes a straight line parallel to and at a distance of B_{og} from the abscissa.

III. Noise Measure

A. An Expression for the Noise Measure in Terms of Two-port Parameters and Source Admittance

Substituting (1) and (9) into (2), the noise measure can be expressed in terms of the gain parameters, noise parameters, and source admittance as follows.

$$M = \frac{F_{min} - 1 + \frac{R_{ef}}{G_s}[(G_s - G_{of})^2 + (B_s - B_{of})^2]}{1 - \frac{1}{G_{a\,max}} - \frac{R_{eg}}{G_s}[(G_s - G_{og})^2 + (B_s - B_{og})^2]}. \quad (15)$$

Rearranging (15), the following equation can be derived,

$$(G_s - G_M)^2 + (B_s - B_M)^2 = G_{RM}^2 \quad (16)$$

where

$$G_M = \frac{M\left(2R_{eg}G_{og} - \frac{1}{G_{a\,max}} + 1\right) + 2R_{ef}G_{of} - F_{min} + 1}{2(MR_{eg} + R_{ef})} \quad (17)$$

$$B_M = \frac{MR_{eg}B_{og} + B_{of}R_{ef}}{MR_{eg} + R_{ef}} \quad (18)$$

$$G_{RM} = \frac{1}{2(MR_{eg} + R_{ef})}\left[\left\{M\left(\frac{1}{G_{a\,max}} - 1\right) + F_{min} - 1\right\}^2 \right.$$
$$+ 4(MR_{eg}G_{og} + R_{ef}G_{of})\left\{M\left(\frac{1}{G_{a\,max}} - 1\right) + F_{min} - 1\right\}$$
$$\left. - 4MR_{eg}R_{ef}(|Y_{og}|^2 + |Y_{of}|^2 - 2G_{og}G_{of} - 2B_{og}B_{of})\right]^{1/2} \quad (19)$$

For the condition

$$G_{RM} = 0, \quad (20)$$

M takes on its minimum value M_{min}, i.e.,

$$M_{min} = \frac{M_2}{M_1}\left[1 + \left(1 - \frac{M_1 M_3}{M_2^2}\right)^{1/2}\right] \quad (21)$$

where

$$M_1 = \left(1 - \frac{1}{G_{a\,max}}\right)^2 + 4\left(1 - \frac{1}{G_{a\,max}}\right)R_{eg}G_{og} \quad (22)$$

$$M_2 = \left(1 - \frac{1}{G_{a\,max}} + 2R_{eg}G_{og}\right)(F_{min} - 1 - 2R_{ef}G_{of})$$
$$+ 2R_{eg}R_{ef}(|Y_{og}|^2 + |Y_{of}|^2 - 2B_{og}B_{of}) \quad (23)$$

$$M_3 = (F_{min} - 1)^2 - 4(F_{min} - 1)R_{ef}G_{of}. \quad (24)$$

The source conductance and susceptance which produce the minimum noise measure are

$$G_{om} = \frac{M_{min}\left(2R_{eg}G_{og} - \frac{1}{G_{a\,max}} + 1\right) + 2R_{ef}G_{fo} - F_{min} + 1}{2(M_{min}R_{eg} + R_{ef})} \quad (25)$$

$$B_{om} = \frac{M_{min}R_{eg}B_{og} + R_{ef}B_{of}}{M_{min}R_{eg} + R_{ef}}. \quad (26)$$

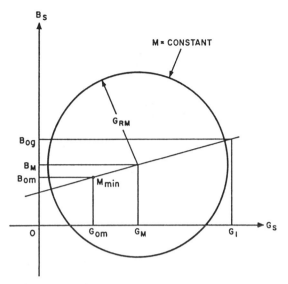

Fig. 3. Constant noise measure locus on rectangular source admittance plane.

B. A Graphical Representation of Noise Measure on the Rectangular Source Admittance Plane

Equation (16) represents a family of circles on the rectangular coordinates with G_s as the abscissa and B_s as the ordinate as shown in Fig. 3. Each circle gives a constant M locus and has its center at the point (G_M, B_M) and a radius of G_{RM}. Since G_M, B_M, and G_{RM} are functions of M, the locus of the center points (G_M, B_M) is not parallel to the abscissa. However, this locus is still given by a straight line through points (G_{om}, B_{om}) and (G_s, B_{og}), corresponding to unity G_a.

IV. Bilinear Transformations of the Graphical Representations

Equations (11) and (16) and a corresponding equation for the noise figure can be rewritten in a form more convenient for representing the behavior of G_a, F, and M on the complex reflection coefficient plane [6], defined as

$$\rho = u + jv = |\rho| \exp(-j\varphi) = \frac{1 - y_s}{1 + y_s} \quad (27)$$

where

$$y_s = g_s + jb_s = \frac{1}{Y_0}(G_s + jB_s) \quad (28)$$

and Y_0 is the (real) characteristic admittance of the input transmission line. A merit of this plane is that a half-plane with either positive or negative conductance can be represented entirely within a unit circle.

Since ρ is a bilinear function of the complex number y_s, any circle in the Y_s-plane can be transformed into a circle in the ρ-plane. Thus, (11) and (16) and the corresponding equation for the noise figure can be expressed in the form

$$(u - u_i)^2 + (v - v_i)^2 = r_i^2 \quad (29)$$

where the subscript j represents g, f, and m for G_a, F, and M, respectively. (u_i, v_i) and r_i, respectively, give the coordinate points for the center of a constant G_a (or F, M) circle and its radius. The center point is also given by the polar coordinates, $|\rho|$ and φ, in the ρ-plane. Furthermore, the Smith-chart coordinates, g_s and b_s, are related to the rectangular coordinates in the manner [7],

$$g_s = \frac{1 - u^2 - v^2}{(1 + u)^2 + v^2} \quad (30)$$

$$b_s = \frac{-2v}{(1 + u)^2 + v^2}. \quad (31)$$

Therefore, the center point can also be specified in terms of g_s and b_s in the same ρ-plane.

Since φ is constant for all constant G_a and F loci as seen later, the polar coordinates is most convenient for mapping G_a and F in the ρ-plane. On the other hand, φ depends upon M so that a preferable coordinate is no longer offered for drawing constant M loci.

In the polar ρ-plane the center point and the radius of a constant G_a circle are given as follows:

$$|\rho_g| = \frac{[(1 - g_{og}^2 - b_{og}^2)^2 + 4b_{og}^2]^{1/2}}{(1 + g_{og})^2 + b_{og}^2 + 2\delta_g} \quad (32)$$

$$\varphi_g = \tan^{-1}\left[\frac{2b_{og}}{1 - g_{og}^2 - b_{og}^2}\right] \quad (33)$$

$$r_g = \frac{2g_{RG}}{(1 + g_{og})^2 + b_{og}^2 + 2\delta_g} \quad (34)$$

where

$$\delta_g = \frac{1}{2R_{eg}Y_0}\left(\frac{1}{G_a} - \frac{1}{G_{a_{max}}}\right) \quad (35)$$

and g_{og}, b_{og}, and g_{RG} are normalized values of G_{og}, B_{og}, and G_{RG} with respect to Y_0. The radial coordinate $|\rho_{og}|$ for $G_{a_{max}}$ is then given by taking δ_g equal to zero. Figure 4 illustrates the geometrical relations of the above parameters in the ρ-plane.

For the noise figure, completely identical forms to the above are available simply by replacing the subscripts. In this case δ_f should be

$$\delta_f = \frac{F - F_{\min}}{2R_{ef}Y_0}. \quad (36)$$

If the Smith-chart coordinates are employed for the mapping of the noise measure, the center point and the radius of a constant M circle are given by

$$g_m = \frac{(1 + g_M)[(1 + g_M)g_M + b_M^2 - g_{RM}^2] - b_M^2}{(1 + g_M)^2 + b_M^2} \quad (37)$$

$$b_m = \frac{b_M[(1 + g_M)^2 + b_M^2 - g_{RM}^2]}{(1 + g_M)^2 + b_M^2} \quad (38)$$

$$r_m = \frac{g_{RM}}{(1 + g_M)^2 + b_M^2 - g_{RM}^2} \quad (39)$$

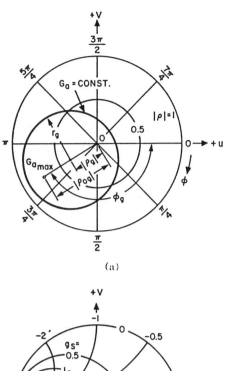

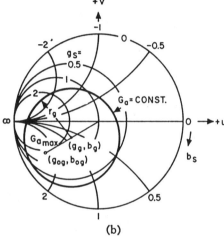

Fig. 4. (a) Constant available power gain locus on complex reflection coefficient plane. (b) Constant available power gain locus on the Smith-chart.

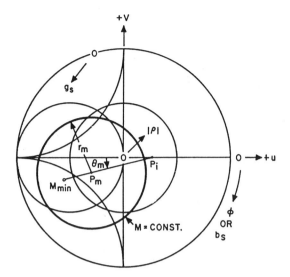

M_{min}: $(|\rho_{om}|, \phi_{om})$ OR (g_{om}, b_{om})
P_i: $(|u_i|, 0)$ OR $(g_i, 0)$
P_m: $(|\rho_m|, \phi_m)$ OR (g_m, b_m)

Fig. 5. Constant noise measure locus on the Smith-chart.

where again small letters indicate normalized values of capital letters with respect to Y_0. The coordinates for M_{min} are, of course, designated by g_{om} and b_{om}.

Due to the nature of the bilinear transformation, the locus connecting the centers of the constant-M circles is still a straight line on the ρ-plane. Its angle θ_m and intersecting point g_i with the abscissa are given by

$$\tan \theta_m = \left[\left\{ F_{min} - 1 - R_{ef}Y_0\left(\frac{|Y_{of}|^2}{Y_0^2} + \frac{2G_{of}}{Y_0} + 1\right) \right\} R_{eg}B_{og} \right.$$
$$- \left\{ \frac{1}{G_{a_{max}}} - 1 - R_{eg}Y_0\left(\frac{|Y_{og}|^2}{Y_0^2} + \frac{2G_{og}}{Y_0} + 1\right) \right\} R_{ef}B_{of} \right]$$
$$\cdot \left[(F_{min} - 1)\left(\frac{|Y_{og}|^2}{Y_0^2} - 1\right)\frac{R_{eg}Y_0}{2} - \left(\frac{1}{G_{a_{max}}} - 1\right) \right.$$
$$\cdot \left(\frac{|Y_{of}|^2}{Y_0^2} - 1\right)\frac{R_{ef}Y_0}{2} + \left\{ |Y_{of}|^2\left(\frac{G_{og}}{Y_0} + 1\right) \right.$$
$$\left. - |Y_{og}|^2\left(\frac{G_{of}}{Y_0} + 1\right) + Y_0(G_{of} - G_{og}) \right\} R_{eg}R_{ef} \right]^{-1} \quad (40)$$

$$g_i = \frac{(1 + g_{om})g_{om} - (\cot \theta_m - b_{om})b_{om}}{1 + g_{om} + b_{om} \cot \theta_m}. \quad (41)$$

θ_m can be determined only by the gain and the noise parameters and is independent of M as seen in (40). For the mapping of constant M circles, the center points can also be given by this line and either g_m or b_m, instead of g_m and b_m. In practice, this line is simply drawn by connecting two representative points, (g_{om}, b_{om}) and (g_1, b_{oo}), corresponding to unity G_a. Figure 5 shows the mapping of constant M locus on the Smith-chart where the polar reflection coefficient coordinates also overlap.

V. Exchangeable Power Gain, Extended Noise Figure, and Extended Noise Measure

The noise figure is normally defined in terms of available power gain as given in (1). However, the available power concept leads to difficulties when the input and output admittances of two-ports have negative real parts. Such cases can, in general, be handled in a similar manner by using the exchangeable power and gain concepts in place of available power and gain. This leads to an extended definition of the noise figure [8]. Using the exchangeable power gain G_e and the extended noise figure F_e, an extended noise measure M_e can be defined as [3]

$$M_e = \frac{F_e - 1}{1 - \frac{1}{G_e}}. \quad (42)$$

In the above cases, the graphical representations of G_e, F_e, and M_e on the source admittance planes are also available in similar ways. In some cases, minor modifications will be required, for example, use of a negative conductance Smith-chart [9], an expanded Smith-chart outside the unit circle to include both positive and negative conductances, and so on.

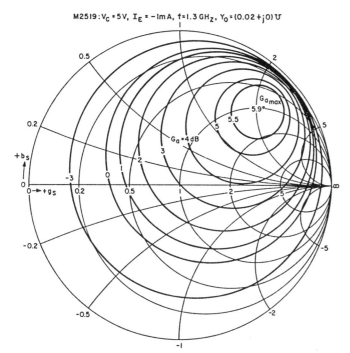

Fig. 6. Available power gain chart.

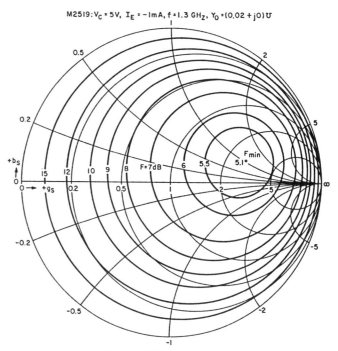

Fig. 7. Noise figure chart.

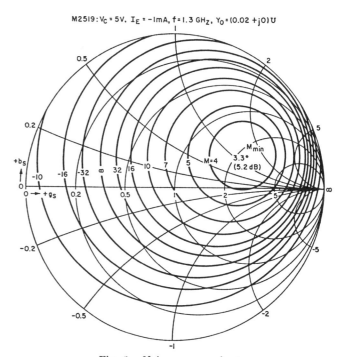

Fig. 8. Noise measure chart.

VI. Examples of Graphical Representations

A. Device Characteristics

A silicon *n-p-n* double-diffused planar microwave transistor M2519 was used in actual representations of available power gain, noise figure, and noise measure of its equivalent two-port. This transistor was characterized at an emitter current of -1 mA, collector voltage of 5 volts, and frequency of 1.3 GHz. The measurements of the gain and the noise parameters were made using the method recommended by the IRE Standard Committee on Electron Tubes [1], [10]. The gain parameters can also be determined by a rather simple method in which $G_{a_{max}}$, G_{og}, and B_{og} are directly obtained. Using the relation of (12), R_{eg} can then be determined by the aid of another measurement of the available power gain with a given source admittance, for example, $Y_s = (0.02 + j0)$ mho. Another way for determining the gain parameters is to use the measured *y*-parameters (or any suitable two-port parameters) as seen in (5) through (8) and (10).

Predetermined parameter values are as follows:

$G_{a_{max}} = 3.93 (5.9 \text{ dB})$
$R_{eg} = 2.54 \text{ ohms}$
$G_{og} = 18.4 \text{ millimhos}$
$B_{og} = 44.2 \text{ millimhos}$
$F_{min} = 3.25 (5.1 \text{ dB})$
$R_{ef} = 15.6 \text{ ohms}$
$G_{of} = 53 \text{ millimhos}$
$B_{of} = 20 \text{ millimhos}$

From the above parameters, the following parameters were obtained:

$M_{min} = 3.32 (5.2 \text{ dB})$
$G_{om} = 47 \text{ millimhos}$
$B_{om} = 30 \text{ millimhos}$

B. Graphical Representations on the Smith-Chart

The available power gain, noise figure, and noise measure for the same transistor are shown on a Smith-chart representation of source admittance, as shown in Figs. 6, 7, and 8, respectively. From these figures, it can be understood easily and precisely what source admittance is able to give a good performance for the amplifier and how the situation becomes worse as the source admittance differs from the optimum value.

As expected, the locus for infinite M coincides with

that for unity G_a and the coordinate point for M_{min} is located between those for $G_{a_{max}}$ and F_{min}.

VII. Conclusions

In this paper it has been shown that the noise measure of linear two-ports can be expressed in terms of the gain parameters, the noise parameters, and the source admittance. Constant M loci have been represented in the source admittance plane as well as constant G_a and constant F loci. Transformations of these loci into the Smith-chart have been made, since this is a very convenient way of representing a wide range of immittances. Such a mapping is useful when the gain and noise performance of amplifiers is investigated, especially in the microwave region. By looking at the noise measure chart, for example, it is easy to deduce what source admittance gives the best overall noise performance and how it deteriorates as the admittance differs from its optimum value. These effects were demonstrated using a microwave transistor as an example.

References

[1] "IRE Standards on Electron Tubes: Methods of Testing," 1962, 62 IRE 7 S1, pt. 9: "Noise in linear twoports."
[2] H. Rothe and W. Dahlke, "Theory of noisy fourpoles," *Proc. IRE*, vol. 44, pp. 811–818, June 1956.
[3] H. A. Haus and R. B. Adler, *Circuit Theory of Linear Noisy Networks*. New York: Wiley, 1959.
[4] ——, "Invariants of linear noisy networks," *1956 IRE Conv. Rec.*, pt. 2, pp. 53–67, 1956.
[5] J. G. Linvill and J. F. Gibbons, *Transistors and Active Circuits*, New York: McGraw-Hill, 1961. The authors gave a wrong expression for available power gain in their book, probably due to typographical errors.
[6] J. C. Slater, "Microwave Electronics," *Rev. Mod. Phys.*, vol. 18, pp. 441–512, October 1946.
[7] P. H. Smith, "Transmission-line calculator," *Electronics*, vol. 12, pp. 29–31, January 1939; and "An improved transmission-line calculator," *Electronics*, vol. 17, p. 130, January 1944.
[8] H. A. Haus and R. B. Adler, "An extension of the noise figure definition," *Proc. IRE (Correspondence)*, vol. 45, pp. 690–691, May 1957.
[9] H. Fukui, "The characteristics of Esaki diodes at microwave frequencies," *1961 ISSCC Digest of Tech. Papers*, pp. 16–17, 1961; and *J. IECE (Japan)*, vol. 43, pp. 1351–1356, November 1960 (in Japanese).
[10] ——, "The noise performance of microwave transistors," *IEEE Trans. on Electron Devices*, vol. ED-13, pp. 329–341, March 1966.

Part II: Noise In Electronic Amplifier Circuits
The Design of Low-Noise Amplifiers

YISHAY NETZER, MEMBER, IEEE

Abstract—The essential theory and practical considerations for the design of low-noise amplifiers are gathered and organized to a uniform presentation. The relevant material is quite simple and straightforward, hopefully bringing within the reach of the interested circuit designer the "art" of low-noise-amplifier design.

I. INTRODUCTION AND SOME SIGNIFICANT CONCLUSIONS

DESPITE numerous papers published on the subject of noise in electronic devices, circuits, and networks in general, the design of low-noise amplifiers is still often regarded by circuit designers as obscure and esoteric. The reason for this seems to be, at least in part, due to the fact that information on actual design of low-noise amplifiers is widely scattered. Most of the published materials are either of theoretical nature—which often tend to discourage the reader, or too superficial for a serious low-noise amplifier design. However, the fact is that circuit design of low-noise amplifiers requires no special knowledge in semiconductor physics, network theory, or probability theory.

Bearing in mind that electronics is a practical science this paper aims to provide a guide and reference source for designing low-noise amplifiers. A comparison is made between the junction transistor, field-effect transistor (FET), and monolithic amplifiers in terms of noise characteristics and their dependence on the bias point, device parameters, and frequency. Noise is treated in terms of the equivalent input noise sources rather than by noise figure, which is less efficient and often confusing. Similarly, the techniques of noise matching are regarded as modifying the input noise sources in such a way that the noise for a given source impedance is minimized.

Some conclusions of practical significance follow.

1) The noise performance of amplifiers, besides being dependent on the amplifier, is also a function of the signal source impedance and frequency range. These two factors determine the optimum input stage.

2) Impedance matching at the amplifier input and source matching techniques for best noise performance are entirely different.

3) For narrow-band reactive sources, the total noise can be reduced by the addition of a suitable reactance at the amplifier input.

4) The noise performance of FET's at low frequencies is related to g_m, and at high frequencies to f_T. Low-noise junction transistors should have a high current gain - β, a minimum base resistance - r_b', and high cutoff frequency f_T.

5) Noise performance of monolithic amplifiers is usually inferior to that of discrete amplifiers. However, mainly at low frequencies, monolithic amplifiers may prove sufficient and should be considered first.

6) For resistive signal sources and low-frequency inductive sensors, "ideal" amplifiers can be designed where the added noise is negligible in comparison to the inherent thermal noise generated in the source.

7) Input devices are now available with $1/f$ noise component reduced to an amount which is virtually insignificant for any practical purpose.

8) High precision in noise calculations serves little purpose, not only because of manufacturing spread of the parameters but also because noise sources are nearly always uncorrelated. As a result, secondary noise sources, such as second stage noise, should have only minor effect.

9) Common-base (or common-gate) input stage has the same input-noise sources as a common-emitter (or common-source) stage. The total performance is inferior, however, except for source impedance of the order of $1/g_m$, especially at high frequencies.

II. SOME NOISE CHARACTERISTICS

Noise can be considered as anything which, when added to the signal, reduces its information content. Here, we shall deal only with noise generated in amplifiers as a result of physical processes occurring in electronic components. This noise is random and, mostly, Gaussian.

Noise cannot be predicted as a function of time. It can, however, be characterized in terms of average values. The most widely used characteristic is the root mean square (rms). For a noise $n(t)$ the rms is defined as follows:

$$\sqrt{\overline{n^2(t)}} = \sqrt{\frac{1}{T}\int_0^T n^2(t)\,dt} \qquad (1)$$

where the bar designates an average value for a relatively long time T.

When two noise sources $n_1(t)$ and $n_2(t)$ are summed, the instantaneous output is the sum of the individual instantaneous values. The *average* output power would be

$$\overline{[n_1(t)+n_2(t)]^2} = \overline{n_1^2(t)} + \overline{2n_1(t)n_2(t)} + \overline{n_2^2(t)}$$
$$= \overline{n_1^2(t)} + 2\gamma\sqrt{\overline{n_1^2(t)}}\sqrt{\overline{n_2^2(t)}} + \overline{n_2^2(t)}.$$

The second term, which is proportional to the average product of the two sources, is a measure of their correlation. γ is defined as the normalized correlation coefficient and its absolute value may vary from 1 to 0. On one extreme, the two sources are identical and differ only in amplitude. On the other extreme, the two sources are totally uncorrelated, and the second term is zero. For zero correlation, the rms of the sum $n_1(t) + n_2(t)$ would be the square root of the sum of the squared individual rms values. In other words, uncorrelated noise adds as orthogonal vectors. For other values of correlation coefficient, the angle between the vectors differs from $90°$. Thus, for example, the addition of a 1-mV noise source to a 2-mV uncorrelated source will increase the latter $\sqrt{5}/2$ times or about 10 percent. Thus one noise source usually dominates and efforts to reduce secondary noise sources are wasteful.

An ideal Gaussian-distribution noise can assume any amplitude as a function of time. In practice, however, even disregarding dynamic range and bandwidth limitations, the percentage of the time during which the instantaneous value exceeds a given amplitude sharply diminishes as a function of this value. For example, the instantaneous noise will exceed an amplitude corresponding to 3.3 times the rms value only 0.1 percent of the time. Similarly, it will exceed an amplitude of 2.5 of the rms 1 percent of the time. Practically, rms values can be measured rather accurately by displaying the noise on an oscilloscope display. The observed noise "thickness" would, roughly, be five times the rms. This method can obviate special measurement equipment and can be refined by overlapping *two* traces. Accuracies on the order of 10 percent can thus be achieved [1]. Another advantage of measuring noise with an oscilloscope rather than with an rms meter or a calibrated average reading ac meter is that power supply ripple or externally induced interferences are not mistaken as "true" noise.

Random noise can also be characterized in the frequency domain. The most important characteristic is the spectral density function (SDF) which is defined as the Fourier transform of the temporal autocorrelation function [2]. Practically, it represents the time averaged noise power $\overline{n^2(t)}$ over a 1-Hz bandwidth as a function of frequency. If the SDF is independent of frequency in the range of interest, it would be referred to as "white noise." The SDF of a voltage $e_n(t)$ is designated $\overline{e_n^2}(f)$, and of a current $I_n(t)$ by $\overline{I_n^2}(f)$. For short, the designations $\overline{I_n^2}$ and $\overline{e_n^2}$ will be used. If $\overline{e_n^2}$ and $\overline{I_n^2}$ are correlated, a normalized correlation coefficient can be defined as for the temporal case. Now, however, it can attain a complex value as well. The meaning of an imaginary component of a correlation coefficient is that a component of one noise can be obtained from the other one by a suitable phase-shifting network, mostly by differentiation or integration. Fortunately, however, the correlation coefficients rarely have any significance in practical low-noise design.

Noise sources are often characterized in terms of spot value, defined as the square root of the power density designated as $V(rms)/\sqrt{Hz}$ or $A(rms)/\sqrt{Hz}$. Practically convenient, but lacking intuitive physical meaning, the spot noise is numerically the rms value over a 1-Hz bandwidth. Due to the similarity to vector quantities as for the temporal definitions, we shall use vector notation for noise sources defined as $\vec{e_n} = \sqrt{\overline{e_n^2}}$ and $\vec{I_n} = \sqrt{\overline{I_n^2}}$.

To calculate the rms value of *white* noise over a frequency bandwidth B, its spot value should be multiplied by $\sqrt{B}$. This is because the noise at different frequency bands is uncorrelated. For similar reasons, the rms value of white noise passing through a network with a transfer function $H(j\omega)$ will be proportional to

$$\sqrt{\int_0^\infty |H(j\omega)|^2 \, d\omega}.$$

If, for example

$$H(j\omega) = \frac{1}{1 + j\omega/\omega_0}$$

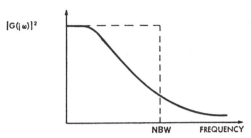

Fig. 1. Noise bandwidth of a low-pass amplifier.

and the input noise has a uniform density of $K(V(rms)/\sqrt{Hz})$, the overall noise power at the output will be

$$\int_0^\infty |H(j\omega)|^2 K^2 \, d\omega = \int_0^\infty \frac{K^2}{1 + (\omega/\omega_0)^2} \cdot d\omega$$

$$= \frac{\pi}{2} \omega_0 K^2 (V^2).$$

In other words, the total noise output is the same as to that of an ideal low-pass filter of bandwidth $\pi/2\,\omega_0$. This bandwidth (see Fig. 1) is referred to as the noise bandwidth of the filter, and accounts for the noise passed beyond the 3-dB frequency, which is usually used to define cutoff. Thus it serves as a correction factor but will assume different values for other than white noise.

The noise bandwidth of an amplifier or network with a known gain can, in principle at least, be determined by supplying a calibrated white noise to the input and measuring the output rms noise. The simplest way to generate such noise is by means of a solid-state noise diode (specially processed Zener diodes) which may give a flat noise density spanning the range $10–10^7$ Hz, a typical noise density of the CND6000-series noise diodes made by Standard-Reference Labs. Inc., is 0.05 $\mu V/\sqrt{Hz}$.

The spectral density of a voltage developed on some complex impedance $Z(s)$ at a particular frequency ω due to the passage of a current noise $\overline{I_n^2}$ would be $|Z(j\omega)|^2 \overline{I_n^2}$. Over a finite bandwidth the rms value would be

$$\sqrt{\int_{\omega_1}^{\omega_2} |z(j\omega)|^2 \overline{I_n^2} \, d\omega}.$$

III. Noise Sources in Electronic Devices

Noise in electronic devices can be attributed to two main processes: thermal noise and shot noise. As a result of thermal fluctuations of charge carriers a noise voltage can be measured in series with a resistor of a value R, whose power density is

$$\overline{e_n^2} = 4kTR \, (V^2/Hz) \quad \text{(Johnson formula)} \tag{2}$$

where T is the absolute temperature of the resistor and $k = 1.38 \times 10^{-23}$ J/K is Boltzmann constant. This density is constant to frequencies up in the infrared, where it begins to drop due to quantum-mechanical effects. It follows, an actual resistor can be represented by a noiseless resistor in series with a voltage noise source. Or, equivalently, by a parallel current noise source $\overline{I_n^2} = 4kTG \, (A^2/Hz)$, where $G = 1/R$ as shown in Fig. 2.

For room temperature (300 K), substitution of the constants yields convenient approximate expressions. So, if R is the resistance in kilohms, the corresponding noise voltage is ap-

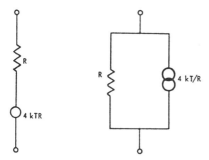

Fig. 2. Thermal noise in resistors.

proximately

$$4\sqrt{R} \text{ (nV/}\sqrt{\text{Hz}}\text{)} \tag{2-a}$$

whereas, the equivalent current source is $4/\sqrt{R}$ (pA/$\sqrt{\text{Hz}}$). For example, the noise voltage associated with a 9-k resistor at room temperature would be $\vec{e_n} = 4\sqrt{9} = 12$ nV/$\sqrt{\text{Hz}}$ and the noise current $\vec{I_n} = 4/\sqrt{9} \sim 1.3$ pA/$\sqrt{\text{Hz}}$. Across a 10-kHz bandwidth the noise voltage will be $12\sqrt{10^4} = 1.2$ µV (rms) and the noise current $1.3\sqrt{10^4} = 130$ pA (rms).

The generation of thermal noise is ideally not affected by the flow of current through the resistor. However, carbon resistors, in particular, have an additional current dependent noise which makes them unsuitable for critical applications. In contrast to resistors, ideal capacitors and inductors do not generate noise. For complex impedance $Z(\omega) = R(\omega) + jX(\omega)$ the spot value of the thermal noise will be due to resistor $R(\omega)$, and its density will, thus, be frequency dependent.

Since electric current is composed of discrete charge carriers, fluctuations are present in the current crossing a barrier where the charge carriers pass independently of one another. Examples are: the p-n junction diode where the passage takes place by diffusion; a vacuum-tube cathode where electron emission occurs as a result of thermal motion, and photodiodes where the absorption of photons is involved. This effect does not exist, for example, in metallic conductors because of long-range correlation between charge carriers. The fluctuations manifest themselves as a noise component named shot noise, which can be represented by an appropriate current source in parallel with the dynamic impedance of the barrier across which it is generated. The spectral density of this source is given by the expression

$$\overline{I_n^2} = 2qI_0 \text{ (A}^2\text{/Hz)} \quad \text{(Schottky formula)} \tag{3}$$

where $q = 1.6 \times 10^{-19}$ C is the electron charge, and I_0 the dc current. The above expression applies up to frequencies close to $1/\tau$, where τ is the transit time through the barrier and thus throughout the useful frequency range of any device.

Photoconductive detectors produce generation–recombination (GR) noise in response to a steady irradiance. Hole–electron pairs are generated randomly and recombine randomly by a statistically unrelated process. Thus full GR noise neglecting that which is thermally generated, corresponds to twice shot noise on the absorbed background photon rate for intrinsic photoconductors. As the dc bias is increased, a voltage is reached at which the minority carrier (hole) transit time is less than the lifetime. At this bias, the carriers are swept out of the device before they recombine and the GR noise approaches shot noise on the photocurrent.

For practical calculations it is convenient to substitute for the constants of formula (3). The current spot noise result-

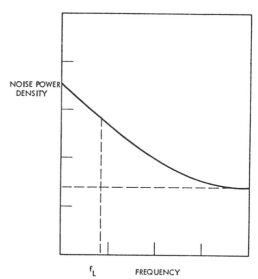

Fig. 3. Flicker noise representation.

ing from I_0 expressed in microampere is:

$$0.57\sqrt{I_0} \text{ (pA/}\sqrt{\text{Hz}}\text{)}. \tag{3-a}$$

For example, the current spot noise associated with an average current of 100 µA, would be $0.57\sqrt{100} = 5.7$ pA/$\sqrt{\text{Hz}}$; across a 10-kHz bandwidth it will amount to $5.7\sqrt{10^4} = 0.57$ nA (rms).

As noted above, thermal noise and shot noise have a constant spectral density. Semiconductor devices as well as vacuum-tubes show, an additional noise component that is inversely proportional to frequency. Hence, it has the name $1/f$ noise (also referred to as excess, flicker, or pink noise) (see Fig. 3). This noise in semiconductors is associated, mainly, with crystal surface conditions. It occurs, however, in nonelectrical phenomena, as well [3]. When associated with current noise, it can be described by the spectral density

$$\overline{I_n^2} = \overline{I_{n0}^2} (1 + f_L/f^n) \tag{4}$$

$\overline{I_{n0}^2}$ represents the white shot noise component, in the excess noise component f_L is the empirical value of the break frequency where the two noise components are equal and is usually subject to process spreads. The actual value of n is not necessarily fixed with frequency. However, in junction transistors it is in the vicinity of 1.1. In some operational amplifiers this value was verified down to 10^{-7} Hz—a one year period! [4]. Apparently, this may lead to very high noise amplitude. However, for the type of spectral density in (4) the noise power in each frequency *decade* is approximately constant. Thus, for example, if f_L is 1 kHz the noise power in the 10^{-7} to 10^3 frequency band is approximately equal to the noise power in the 1- to 30-kHz band. In practice, flicker noise is often regarded as a dc instability.

IV. THE CHARACTERIZATION OF NOISE IN AMPLIFIERS BY EQUIVALENT INPUT SOURCES

As a result of the mechanisms described above, the output of any real amplifier is accompanied with a noise which depends on the measurement bandwidth, the overall gain and the noise properties of the various stages. The sensitivity of the amplifier is best characterized by the minimum signal at the input, still detectable at the output—rather than by the actual

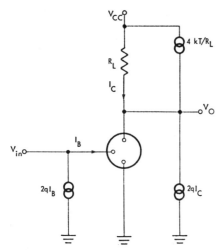

Fig. 4. Noise sources in a three-terminal amplifying device.

noise measured at the output. This input signal may conveniently be defined as equal to a virtual noise source located at the input, which is obtained by dividing the actual output noise over a given bandwidth by the overall gain. As shown later, the total equivalent input noise of a real amplifier also depends on the impedance of the signal source besides thermal noise that may accompany this impedance.

The device shown in Fig. 4 may represent a junction transistor, FET, or vacuum-tube gain stage in a somewhat simplified manner, and can be considered to have the following noise sources:

1) shot noise which accompanies the bias current I_B of the control electrode (base, gate, or grid) and is given by $2qI_B$ A²/Hz;
2) shot noise of the quiescent current I_C, given by $\lambda q I_C$ A²/Hz; λ is dependent upon the particular device, in the junction transistor; it is equal to 2, that is, a "full" shot noise;
3) thermal current noise of the load resistor R_L, given by $4kT/R_L$ (A²/Hz).

In relating these noise sources to the input of the stage, the first source, being already located at the input, can be represented by a current source shunting the input. The second source can be represented by a noise voltage source $\overline{e_n^2}$ in series with the input and given by

$$\overline{e_n^2} = \frac{\lambda q I_C}{g_m^2} \left(\frac{V^2}{Hz}\right).$$

In these devices, the mutual conductance g_m increases with the quiescent current I_C [5], although not necessarily linearly and the actual noise is also found to be proportional to absolute temperature T. If we assume $g_m \sim I_C$ then

$$\overline{e_n^2} \sim \frac{qT}{g_m} \left(\frac{V^2}{Hz}\right).$$

Thus the noise may alternatively be attributed to a thermal origin in $1/g_m$.

The power density of this voltage noise is thus inversely proportional to the mutual conductance of the device. Assuming that the dc voltage on the resistor R_L is V_L, then the shot noise density of I_C would at most be $2q V_L/R_L$ compared to $4kT/R_L$ of the resistor thermal noise. A simple calculation shows that if $V_L > 50$ mV, the contribution of the shot noise exceeds that of the thermal noise and the thermal noise of R_L can practically be ignored. A resistive load is usually preferable to an active current source biasing since the latter would add its own shot noise which is comparable to that generated in the active device. However, "quiet" current sources can be obtained, for example, by means of a bipolar transistor with an emitter degenerating resistor. It is apparent that any biasing resistor in parallel with the input will add its thermal noise to the input equivalent current source. Similarly the noise of a resistor R_e in series with the emitting electrode can be accounted for by adding $4KT/R_e g_m^2$ to $\overline{e_n^2}$.

The above simplified model is applicable over a considerable frequency range, and indicates that the noise sources are determined mainly by the input bias current and transconductance. These sources seem to have a constant spectral density and be statistically independent. However, as shown later, at high frequencies the density of these sources increases due to decreasing gain, inversely to the device cutoff frequency. At low frequencies the spectral density increases as a result of excess noise effects.

To calculate the noise contributed by the second stage, we observe that its bias current shot noise $2qI_{B2}$ is adding directly to the shot noise $2qI_{C1}$. Due to I_B being small compared to I_C, it is negligible. Similarly $\overrightarrow{e_{n2}}$ of the second stage should be compared with $\overrightarrow{e_{n1}}$ after dividing by the voltage gain $g_{m1}R_L$. From Section I, a voltage gain of only 2 in the first stage may still be sufficient to render its contribution negligible. Thus first stage dominates the amplifier noise performance, unless for some reason the second stage has an unusually high noise.

The two equivalent input noise sources model applies to any amplifier regardless of the nature of its components [6]. In fact, different combinations of external equivalent noise sources along with their correlation coefficient can be used to represent the actual noise sources within any amplifier. However, the above representation is the most convenient for practical purposes.

V. INFLUENCE OF SIGNAL-SOURCE IMPEDANCE ON THE TOTAL NOISE

Fig. 5 shows schematically an amplifier with input impedance Z_{in} and white input noise sources $\overrightarrow{e_n}$ and $\overrightarrow{I_n}$ which are statistically uncorrelated. If a voltage signal source e_s with internal impedance R_s is applied to the amplifier input, the signal voltage at the input would be

$$e_s \frac{Z_{in}}{Z_{in} + R_s}$$

and the total noise at the input would be

$$(\sqrt{4kTR_s} + \overrightarrow{e_n} + \overrightarrow{I_n} R_s) \frac{Z_{in}}{Z_{in} + R_s}. \quad (5)$$

This does not include any noise, other than thermal, which may be present in the source which for our treatment would be regarded as a signal.

The signal-to-noise ratio (S/N) at the virtual amplifier input (and the actual S/N at the output) will be $e_s/(\sqrt{4kTR_s} + \overrightarrow{e_n} + \overrightarrow{I_n} R_s)$ and is at maximum for low source resistance. The input impedance Z_{in} apparently does not affect the S/N, since by definition, any noise generated at Z_{in} is implicit in $\overrightarrow{I_n}$.

In general, a signal-source can be represented by either a voltage or current source. In practice, however, signal sources

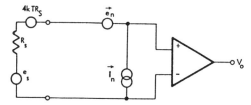

Fig. 5. Noise at the input of an amplifier.

often have one "natural" representation even disregarding their internal impedance. Thus in a current-signal source the short circuit signal current is essentially independent of its internal impedance, as opposed to a voltage-signal source. An analysis similar to the above would show that for best noise performance a current-signal source should have a minimum shunt admittance, even neglecting possible thermal noise and, again, independent of the amplifier input impedance. For a given amplifier, however, an added impedance at the input will adversely affect its performance. It is easy to show, for example, that a shunt parasitic capacitance, such as that associated with a coaxial cable, which may be noiseless by itself, tends to deteriorate the noise performance, especially for current-signal sources at relatively high frequencies.

The above classification does not include parametric sensors, i.e., sensors in which the signal is related to the source impedance, such as infrared resistive detectors. These sensors should be supplied with either current or voltage biasing and the corresponding signal would be accordingly read as a voltage or current. The signal-to-noise ratio is, theoretically at least, proportional to the bias and independent on the actual biasing method, as long as any noise that may be added by the biasing network is negligible.

The total input voltage noise of a given amplifier or device can be determined by measuring the output noise and dividing it by the voltage gain. When the input is shorted, the contribution of output noise is due to $\vec{e_n}$ only. However, when the input is biased with a large enough impedance, the contribution will come mostly from $\vec{I_n}$. If an individual selection of devices for low noise is desired—usually in the low-frequency range, the input noise sources can be measured regardless of the actual gain of the device by a fixture in which the gain is held constant by a feedback network [7]. In this way, individual selection of premium devices can be facilitated.

VI. NOISE MATCHING

The total input noise current of an amplifier with input equivalent sources $\vec{e_n}$ and $\vec{I_n}$, and correlation coefficient γ fed by a signal source with internal complex impedance Z_s is given by

$$\sqrt{\overline{I^2}} = \sqrt{4kT/R_e|Z_s| + \overline{e_n^2}/Z_s^2 + \overline{I_n^2} + 2\gamma \cdot \vec{e_n} \cdot \vec{I_n}/Z_s^2}. \quad (6)$$

This usually determines the minimum signal that can be handled. Nevertheless, this threshold can often be lowered by noise matching. Noise matching is based on the fact that a coupling transformer with turns ratio $1:n$ in series with the amplifier input (Fig. 6(a)) yields an equivalent amplifier with input noise sources modified to $n\overline{I_n^2}$ and $\overline{e_n^2}/n$, the correlation coefficient being unchanged [8].

The input noise now becomes a function of the turns ratio, and reaches a minimum when

$$n^2 = n_{opt}^2 = \frac{1}{Z_s} \cdot \frac{\vec{e_n}}{\vec{I_n}} \quad (7)$$

the corresponding expression for the minimum noise voltage is

$$Z_s [4kT + 2(1 + \gamma)\vec{e_n} \cdot \vec{I_n}]. \quad (8)$$

Thus a step-up transformer is needed, when the voltage noise predominates, and vice versa. If the input noise sources are frequency dependent n_{opt} will be obtained by differentiating the overall noise over the bandwidth with n as a parameter. n_{opt} will be a function of the frequency range but still independent of the correlation coefficient.

From expression (8), the contribution of the amplifier to the input noise is $2Z_s(1 + \gamma)\vec{e_n} \cdot \vec{I_n}$ thus neglecting γ the magnitude $\vec{e_n} \cdot \vec{I_n}$ characterizes the inherent noise performance of an amplifier or an input device, provided noise matching is feasible. In practice, the selection must bring other factors into consideration since coupling by means of a transformer is often incompatible with solid-state circuits, and may involve bandwidth limitations, intrinsic resistance noise, bulkiness, and sensitivity to external magnetic interferences.

In certain situations there is a latitude in the selection of the transducer impedance. For example, in a flux measuring magnetic transducer, such as a reproducing tape head, the desired signal is proportional to the flux in the magnetic core. The short-circuit current, assuming purely inductive impedance, can be shown to be directly proportional to the flux and inversely proportional to the number of turns. However, for any input device the total input current noise at a given bandwidth is also a function of source inductance. Thus there is an optimum number of turns as a function of the magnetic core and the input equivalent noise sources of the amplifier, similar to selecting an optimum turns ratio of a matching transformer. In a sense, the transducer serves as its own matching transformer. In practice, the windings resistance must also be considered and, for the analogy to remain valid, this implies that the wire cross section should be inversely proportional to the number of turns. The total cross section of the winding which is assumed constant should be the maximum possible to minimize source thermal noise.

The reduction of noise with a step-up transformer may in a sense be regarded as due to noiseless voltage amplification prior to the actual amplifier, equivalently reducing the input voltage noise. This may also be effected for a narrow-band resistive source by means of a series resonant circuit, as shown in Fig. 7. At the resonant frequency the overall S/N is identical for the two configurations and is given by

$$\frac{e_s^2}{4KTR_s + \overline{I_n^2} R_s^2 (1 + L/R_s^2 C) + \overline{e_n^2} R_s^2 C/L}$$

as compared to

$$\frac{e_s^2}{4KTR_s + \overline{I_n^2} R_s^2 + \overline{e_n^2}}$$

without the resonant circuit. The two input equivalent noise sources are thus effectively modified in a nearly reciprocal manner, and by properly selecting L and C the narrow-band noise can be minimized. However, as shown below, for a low resistance source the amplifier noise can in most cases be made negligible by merely selecting a proper junction transistor as an input stage.

In cases when $\vec{e_n}$ predominates, noise matching can also be effected by connecting several input devices in parallel. This technique is based on the fact that n identical devices in paral-

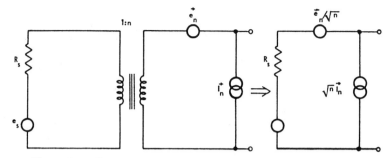

Fig. 6. The effect of input coupling transformer on the equivalent input noise sources.

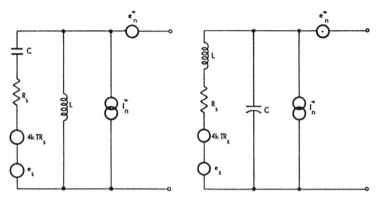

Fig. 7. Input noise sources modification in a resonating resistive signal source.

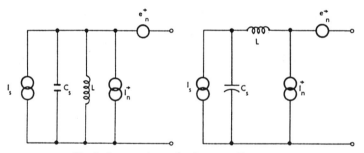

Fig. 8. Input noise tuning for a reactive source.

lel are equivalent, as far as noise sources are concerned, to a single device preceded by an input transformer having the turns ratio $1:\sqrt{n}$ — i.e., the source $\vec{e_n}$ is decreased and the source $\vec{I_n}$ proportionally increases (8). Here, too, the correlation coefficient remains unchanged.

In determining the transformer turns ratio, the number of devices in parallel—or the quiescent current in a bipolar transistor for minimum noise (see below), the expression for the total noise is of form $x + 1/x + c$ where x is the turns ratio, the number of devices in parallel or the emitter current. This function has a shallow minimum and as a result, noise matching is not critical. For example, if x deviates from its optimal value by a factor of 2 the total noise increases by no more than 25 percent.

VII. Noise Reduction for Reactive Sources by Input Tuning

The resonant matching method mentioned above can effectively be applied to inductive narrow-band sources by merely adding proper resonating capacitance. For reactive sources in general, the type of reactance to be applied at the amplifier input depends on the nature of the signal source and the dominant noise source. For a capacitive current signal source I_s, an inductance L in series or in parallel to the amplifier input will have different effects on the S/N as shown in Fig. 8.

The overall current signal at the input of the amplifier may be obtained by shorting B to ground. For a series inductance the signal component in the short-circuit current will be

$$I_s \cdot \frac{1}{1+S^2 LC}$$

whereas the noise component is

$$\frac{\vec{I_n} S^2 LC_s + \vec{I_n} + \vec{e_n} SC_s}{1+S^2 LC}.$$

The signal component is now frequency distorted. After equalization the S/N would be

$$\frac{S}{N} = \frac{I_s}{\vec{e_n} \cdot SC_s + \vec{I_n}(1+S^2 LC)}$$

whereas the S/N without the inductance would have been

$$\frac{S}{N} = \frac{I_s}{\vec{e_n} SC_s + \vec{I_n}}. \qquad (9)$$

The inductance in combination with the source capacitance cancels the source reactance at the resonant frequency and modifies a formerly white current noise $\vec{I_n}$ to one with a spectrum proportional to $|1 - \omega^2 LC|$, disappearing totally at the resonant frequency. However, at higher frequencies the S/N deteriorates compared to $L = 0$. In a similar manner an inductance in parallel to the input yields a S/N

$$\frac{S}{N} = \frac{I_s}{\vec{I_n} + \vec{e_n}(SC_s + 1/SL)}.$$

Now $\vec{e_n}$ vanishes at $\omega = 1/\sqrt{LC}$, however, the S/N deteriorates at low frequencies compared to $L = 0$.

The techniques discussed, basically different from bandpass filtering, can be combined with transformer action to take care of the remaining noise source, subject to practical limitations. The improvement in S/N is inversely proportional to the bandwidth; however, the technique can still be applied to broadband sources. Thus an inductance in series with a broad-band capacitive current source and paralleling several FET's, may decrease the total noise by a factor of 2 [9].

VIII. NOISE FIGURE CHARACTERIZATION OF AMPLIFIERS

An older characterization of amplifier's noise performance is by means of the noise figure (NF). It is defined as the S/N at the amplifier output divided by the corresponding ratio at the input, expressed in decibels

$$NF = 10 \log_{10} \left[\frac{(S/N)_{\text{out}}}{(S/N)_{\text{in}}} \right]. \qquad (10)$$

Equivalently, NF is the log ratio of the total output noise to its portion originating as thermal noise in the signal source resistance. Thus an NF of 3 dB means that half the output noise is due to the amplifier. For many amplifiers and resistive source combinations much lower NF are obtainable. In terms of the input equivalent noise sources NF can be expressed as follows:

$$NF = 10 \log_{10} \frac{4kTR_s + \overline{e_n^2} + 2\gamma \vec{e_n} \vec{I_n} R_s + \overline{I_n^2} R_s^2}{4kTR_s}$$

γ is the correlation coefficient, if any, between the two input noise sources.

Differentiation of the latter expression with respect to R_s yields the so-called optimal source resistance $R_{s\,\text{opt}} = \vec{e_n}/\vec{I_n}$ where NF attains a minimum. In Fig. 9, NF is plotted as a function of R_s for three combinations of $\vec{e_n}$ and $\vec{I_n}$, but with fixed $R_{s\,\text{opt}}$ showing the effect of $\vec{e_n} \cdot \vec{I_n}$ on the function. For given source resistance R_s, NF can be improved by modifying the source resistance to its optimal value, apparently by the addition of a series or shunt resistance. Actually, however, this would lower the S/N at the output because of the added thermal noise prior to amplification. Furthermore, for $R_s = 0$, NF reaches infinity although the actual output noise is less than that corresponding to any other source resistance, including $R_{s\,\text{opt}}$. The "paradox" lies in the fact that the reduction of NF proportionally improves the S/N ratio at the output

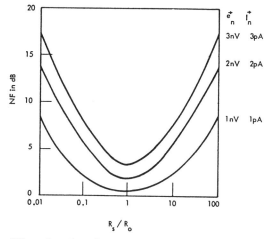

Fig. 9. NF as a function of source resistance for various amplifier figure of merit.

only if the S/N at the source remains unchanged. This can be satisfied by coupling the signal source by means of a transformer. Provided the turns ratio is $\sqrt{R_{s\,\text{opt}}/R_s}$ NF is brought to its minimum, and the S/N to a maximum. This, however, is another interpretation of noise matching as already described.

In general, NF by itself cannot fully characterize the noise performance of an amplifier, nor does it provide a basis for prediction of noise with an arbitrary source impedance. Furthermore, NF does not apply to current signal source or reactive signal sources which ideally have no thermal noise. On the other hand, in RF communications and devices, noise figure is commonly used because of the convenience of optimum matching by a transformer or other source coupling reactive networks.

IX. NOISE IN JUNCTION TRANSISTORS

The equivalent input noise sources of the junction transistor at midfrequencies are given by

$$\overline{I_n^2} = 2qI_B = 2q\frac{I_e}{\beta}$$

$$\overline{e_n^2} = 4KT\left(r_{b'} + \frac{r_e}{2}\right), \qquad r_e = \frac{kT}{qI_e} = \frac{1}{g_m} \qquad (12)$$

where I_B and I_e are the base and emitter currents and $r_{b'}$ and r_e are, respectively, the spreading base resistance and emitter small-signal resistance. The current noise is obviously the shot-noise of the base current, whereas the voltage-noise source corresponds to the thermal noise of the base resistance in series with one-half the small signal resistance of the base-emitter junction. At relatively low emitter currents, when $r_e > r_{b'}$, $\overline{e_n^2}$ is inversely proportional to I_e, whereas $\overline{I_n^2}$ is directly proportional to it as shown in Fig. 10, this means that the operating current can be matched to the signal source impedance in order to minimize the overall noise in much the same way as with a transformer.

The noise sources are independent of the collector voltage as long as the leakage current is negligible and they are essentially the same for various connections [10], however, the common-emitter is usually preferred due to the higher gain. From (12), a transistor with a relatively high β and a low $r_{b'}$ is potentially best for minimum noise. When the source resistance is R_s, the

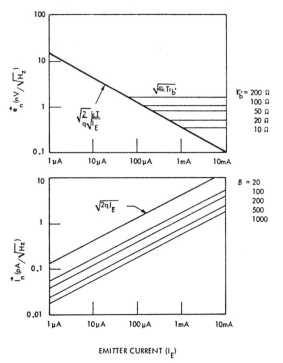

Fig. 10. Graphic representation of junction transistor input noise sources as a function of I_e.

optimum emitter current obtained by differentiation is

$$I_{opt} = \frac{kT}{q} \frac{\sqrt{\beta}}{R_s + r_{b'}}$$

$$= \frac{25\sqrt{\beta}}{R_s + r_{b'}} \quad (mA).$$

For signal sources other than purely resistive, the optimum current will depend on the bandwidth as well. For small source impedance of the order of $r_{b'}$ the input noise sources are no longer reciprocal and the noise performance, and NF, deteriorate. The value of $r_{b'}$ is usually not supplied in data sheets and is subject to manufacturing spread. Typical values vary from several hundred ohms for super-β transistors, to several tens of ohms in certain types [10, pp. 11, 68]. Evidently, $r_{b'}$ can be reduced by a parallel connection of several transistors. This technique is employed in the LM194 matched pair which—not being optimized for low noise, have $r_{b'} = 30\ \Omega$. However, some special geometry transistors have $r_{b'}$ reduced to a few ohms (types 2SD786 and 2SB737,[1] n-p-n and p-n-p transistors with a typical $\beta = 400$ and the specified value of $r_{b'}$ of 4 and 2 Ω, respectively). A junction transistor first stage may thus contribute insignificant noise even with very low resistance sources.

At frequencies where β falls off the input current noise source increases with frequency and is given by

$$\overline{I_n^2} = 2qI_B \left(1 + \beta \frac{f^2}{f_T^2}\right). \quad (12\text{-a})$$

The corner frequency is thus $f_T/\sqrt{\beta}$ where f_T is the cutoff frequency. Similarly, $\overline{e_n^2}$ starts to increase near the upper use-

---

[1] Made by TOYO Electronics Ind. Corp., Central Kyoto, Japan (represented by R-Ohm Corp., P.O. Box 4455, Irvine, CA 92761).

ful frequency range of the transistor, along with some correlation to $\overline{I_n^2}$ [12]. Considering the fact that f_T increases with I_e, the optimal current for a given source impedance will increase at high frequencies reflecting the fact that the effective β is decreasing.

The effect of operating the junction transistor at low temperatures is, apparently, to reduce $\vec{e_n}$ proportionally to T. In silicon planar transistors $\vec{e_n}$ was found to reach a minimum of around 150 K–[13]; however at still lower temperatures, it increases again, accompanied by a sharp drop in β and the cutoff frequency, deteriorating $\vec{I_n}$ as well. A similar trend is found also in germanium transistors.

X. Noise in Field-Effect Transistors

The noise sources of FET's above the excess noise region are given by the following expressions [14], [15]:

$$\overline{e_n^2} = 0.7 \cdot 4kT/g_m$$
$$\overline{I_n^2} = 2qI_g + 0.7 \cdot 4kT/g_m \cdot \omega^2 C_{gs'}^2 \quad (13)$$

g_m is the mutual conductance of the FET, I_g is the gate leakage current, and $C_{gs'}$ is the internal input capacitance of the FET (roughly $\frac{2}{3}$ of the total capacitance C_{gs}). As expected, the noise sources have a form similar to that in the junction transistor. However, the leakage current I_g is usually much smaller than a typical base current. It is also not proportional to the operating current, it may, however, increase substantially at drain voltage above several volts, thus there is no interdependence between the two noise sources. The noise sources of the MOSFET are much the same as those of the junction FET but I_g is negligible. In junction FET's $\overline{e_n^2}$ is found in practice to be somewhat larger than calculated from the measured g_m. This is due to the thermal noise of the bulk resistance of the source terminal, the actual measured $\vec{e_n}$ is usually not less than 2 nV/$\sqrt{Hz}$. Special geometry junction FET's such as the 2N6550 achieve $\vec{e_n} = 0.8$ nV/$\sqrt{Hz}$ by increasing g_m along with C_{gs}.

A special meshed-gate geometry developed by TOSHIBA yields extremely low noise FET's for audio frequencies. In this family the p-channel FET type 2SJ72 and the n-channel type 2SK147 (with its dual version 2SK146) have $\vec{e_n} = 0.75$ nV/$\sqrt{Hz}$ with $C_{gs} = 130$ pF and 50 pF, respectively. This noise level is equivalent to that of a 35-Ω resistor and is achieved at a drain current of 2 mA. For still higher frequencies, the 2SK117 is available with $\vec{e_n} = 1$ nV/$\sqrt{Hz}$ and $C_{gs} = 10$ pF at a drain current of only 0.5 mA.

To reduce $\vec{e_n}$ in FET, g_m apparently, must be increased as much as possible by maximizing drain current I_D. However, because g_m is proportional to $I_D^{1/2}$, $\vec{e_n}$ is proportional to $I_D^{-1/4}$, and with such a mild dependence, there is no advantage in increasing I_D beyond a value dictated by other considerations. In addition, excessive heat dissipation reduces the effective g_m and increases the leakage current I_g. More effectively, noise can be reduced, when necessary, by connecting several FET's in parallel or selecting a large geometry FET's mentioned above, however, due to the corresponding increase of C_{gs}, high frequency noise performance is impaired and there is an optimal number of FET's. For capacitive source, the optimum occurs when the total capacitance at the input becomes roughly equal to that of the source [9], and in general, the noise quality of the FET is dependent on the ratio g_m/C_{gs}, which is also the high-frequency figure of merit. An example of a high-quality FET is U309, with $g_m = 15$ mmho and $C_{gs} = 4.3$ pF. Another

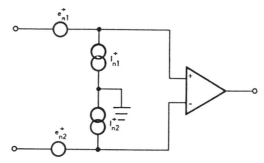

Fig. 11. Noise sources of a differential input stage.

		$\vec{e}_n$ (nV/$\sqrt{Hz}$)	$\vec{e}_n$ Low-Frequency Corner (Hz)	$\vec{i}_n$ (pA/$\sqrt{Hz}$)	$\vec{i}_n$ Low-Frequency Corner	Gain Bandwidth Product (MHz)	Notes
HA-909		7	100	0.2	2.0 kHz	7	
HA-4602		7	300	0.15	1.5 kHz	8	Quad
NE-5534A		4	100	0.4	200 Hz	10	
OP-27/37		3	3	0.4	140 Hz	8/63	
OP-07		10	10	0.1	50 Hz	0.6	
OPA 101/102 BM,	Burr-Brown	8	100	0.002		20/40	Spec. Guaranteed
PM 156/157 A		12	50	0.01	<100 Hz	4.5/20	Bifet
MA-334	Analog-	8	60	0.005	60 Hz	15	
MA-322	Systems	3.5	100	0.5	400 Hz	50	
MA-106		0.6	1000	2		15 (3 dB)	
ZN-459	Ferranti	0.8	50	1		15 (3 dB)	Video-Amplifiers
SL1205C	Plessey	0.8	<100			6.5 (3 dB)	

Fig. 12. Currently available low noise monolithic amplifiers.

high-performance FET is TOSHIBA type 2SK61, this device has $g_m = 10$ mmho, $C_{gs} = 4$ pF, and a very small reverse capacitance $C_{gd} = 0.1$ pF.

The ratio g_m/C_{gs} is still higher in FET's of the D-MOS type, such as Signetics type SD203, with $g_m = 15$ mmho and $C_{gs} = 2.4$ pF. A disadvantage of MOSFET's, in general, is the high level of flicker ($1/f$) noise in $\vec{e}_n$ which precludes its use at the audio range. Compared to silicon FET's gallium-arsenide FET's have higher g_m/C_{gs} ratios and, thus, are potentially lower in noise. However, presently they suffer from very high $1/f$ noise and high gate leakage current. Consequently, they are useful only for very high frequencies.

In silicon junction FET's, g_m usually increases by decreasing temperature, in addition to the implicit effect of T in the expressions for the two noise sources and the leakage gate current. A FET front-end may thus be cooled to advantage in case the signal source is already cooled. It is found, however, that the improvement in $\vec{e}_n$ reaches a maximum around 100 K, depending on the specific device [16]. MOSFET's, on the other hand, can be operated at still lower temperatures [17].

XI. Noise in Monolithic Amplifiers

The noise sources in monolithic amplifiers are essentially those expected from a discrete equivalent, with the first stage usually being of a differential junction or FET pair. Monolithic operational amplifiers, as well as discrete junction transistors, may suffer from additional type of noise, beside those expected from (12). This is called "burst" or "popcorn" noise and is associated with the base current noise (see Section XII). "Burst" noise is not a Gaussian noise but more like random jumps between two levels with characteristic times spanning a large range. However, it is virtually absent in most modern devices due to improved manufacturing technologies.

The differential input amplifier can be represented in terms of four noise sources, two at each input (Fig. 11) similar to dc drift representation. The two voltage noise sources can be vectorially combined and represented by a single equivalent source connected in series with one of the inputs. Then the differential input stage is represented by two current noise sources and one voltage noise source.

Monolithic operational amplifiers have usually been optimized for input dc characteristics such as low bias current high dc gain, and low power consumption rather than for low noise. Moreover, some of the circuit techniques utilized such as use of Darlington input stage, active loads in the first stage or the use of resistive input protection network tend to degrade the noise performance. In addition, high β is accompanied with large $r_{b'}$. In some operational amplifiers the input bias current is internally supplied, with the result that the input current noise is much more than expected due to the specified bias current. The above drawbacks combined with high-frequency limitations and the inability to modify the first stage current for matching various sources have tended to limit the use of operational amplifiers and other monolithic amplifiers in various critical applications. However, operational amplifiers and other monolithic amplifiers in particular, have been made available in recent years specifically for low noise (Fig. 12), and can fit many low-noise applications with the benefits of small space and low price. For comparison, a μA741 amplifier has—depending on the manufacturer—a typical $\vec{e}_n = 25$ nV/$\sqrt{Hz}$ and $\vec{I}_n = 0.6$ pA/$\sqrt{Hz}$. This means that for any source impedance the amplifier noise will exceed the source thermal noise, this is not true for the above amplifiers each within its typical range of source impedance.

In cases where lowest noise is desirable for a given range of source impedance, a discrete differential input stage can be

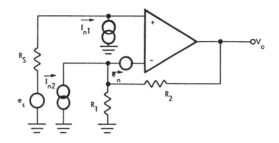

Fig. 13. Noise sources in a noninverting-connection operational amplifier.

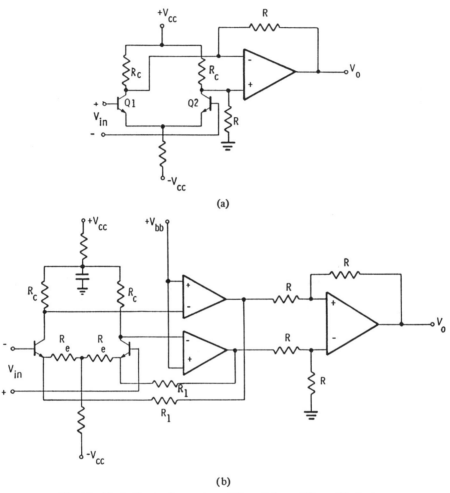

(a)

(b)

Fig. 14. (a) A discrete input stage differential amplifier. (b) An improved discrete input stage differential amplifier.

combined with a monolithic operational amplifier. The input stage being optimized for noise while most of the loop gain is supplied by the operational amplifier. However, the addition of the input stage often makes the amplifier unstable in closed loop without further compensation.

The general expression for the total input noise in a noninverting operational amplifier (Fig. 13) with a resistive source is

$$\overline{e_n^2} + \overline{I_n^2}(R_1//R_2)^2 + 4kT(R_1//R_2) + 4kTR_s + \overline{I_n^2}R_s^2. \tag{14}$$

The value of the feedback network resistors must be low enough to ensure minimum added noise. In comparison the inverting-type configuration is not suitable for low noise due to the series input resistor. For a similar reason the conventional four resistors configuration of a differential amplifier is not suitable for low noise applications. When low noise differential input is needed, the differential input pair can be left outside the feedback loop as in Fig. 14(a) where the low frequency gain is $g_m R$. A more elaborate configuration is shown simplified in Fig. 14(b). In this scheme the input pair is within the feedback loop, yet the signal source can directly coupled to the bases because feedback is applied to the emitters. The differential gain of the amplifier is $2R_1/R_E + 1$, and the common mode as well as power supply rejection ratio can be made very high by selecting matched components. It may be necessary though, to bypass the positive and negative supplies near the input stage to eliminate possible high-frequency supply noise from coupling through unbalanced stray capacitances at the collectors. This is true also for operational amplifier preamplifiers due to the finite power supply rejection ratio.

In general, even though a single-ended signal source would enjoy less added noise when coupled to an amplifier with a single-ended input stage, say a common emitter followed by an operational amplifier, the power supply sensitivity, components count, and size of coupling and bypass capacitors may still make the differential input amplifier a better choice.

In some cases an amplifier must have both low input noise, possibly with wide bandwidth, and good dc properties (offset voltage and input bias current) as well. These requirements tend to conflict; however they can be met using a composite amplifier [18]. Similarly to the classical chopper-stabilized configuration, the amplifier is separated into two channels, one which has a differential input stage and is dc coupled to the signal source, and another channel which is a low-noise ac amplifier. The two output signals are then combined to a single channel. If the dc amplifier output is low passed, its input voltage noise source will not contribute to the output noise above the cutoff frequency. The total input current noise however originates from the two inputs: By design, however, the input bias current of the dc amplifier is small and has a low shot noise. At high frequencies the current noise which flows through the source impedance tends to increase, but its effect can be suppressed by a decoupling choke in series with the input of the dc channel.

XII. Input Device Selection for Very-Low Frequencies

Expressions (12) and (13) give the noise sources in the FET and junction transistors but do not take into account excess noise effects at low frequencies, which are less predictable.

As a general rule, in junction transistors low-frequency excess noise is associated with the current noise source I_n [19], [20]. Another noise associated with the base current but which is usually of no concern in discrete devices is the *burst* noise mentioned earlier [21], [22]. The opposite occurs in the FET and especially MOSFET's where the voltage noise source e_n is affected.

In junction transistors $\vec{I_n}$ is given by

$$\overline{I_n^2} = 2qI_B + \frac{KI_B^m}{f^n} \quad (15)$$

whereas in FET's $\vec{e_n}$ is approximated by

$$\overline{e_n^2} = 4kT\frac{0.7}{g_m}\left(1 + \frac{f_L}{f^n}\right). \quad (16)$$

In the above expressions, $n \sim 1$, K, and f_L are subject to spread whereas $1 < m < 2$ [23]. The low-temperature dependence of low-frequency noise is pronounced and is such that n is not fixed [17], [24].

In FET's, not specified for low noise at low frequencies, the corner frequency f_L may be of the order of many kilohertz; on the other hand, in some N-channel type, f_L can be as low as 1 kHz. These FET's are usually characterized by $\vec{e_n}$ at 10 Hz. Devices such as 2N6483, 2N5592, 2N4867A, 2N6550, and NF 101, may have $\vec{e_n}$ as low as 6 nV/$\sqrt{\text{Hz}}$ which can of course, be further lowered by paralleling. By far, however, the best low-noise junction FET's are TOSHIBA family of meshed-gate devices, mentioned earlier. The corner frequency may be as low as 15-20 Hz and at operating current $I_d = 2$ mA the voltage noise at 10 Hz is 2 nV/$\sqrt{\text{Hz}}$ for the 2SJ72 and 1.3 nV/$\sqrt{\text{Hz}}$ for the 2SK147.

Before the advent of low flicker noise FET's, the only means for obtaining low $\vec{e_n}$ and a negligible $\vec{I_n}$, at low frequencies had been the varactor-bridge amplifier (also referred to as a low-frequency parametric amplifier). This amplifier is based on signal dependent imbalancing of a voltage-controlled capacitor (varactor) bridge, which is driven by a reference high-frequency carrier. The output of the bridge is a carrier modulated by the signal and after being further amplified is synchronously demodulated. The initial gain is achieved through the voltage dependent capacitance of the varactors, rather than by an active device. The source $\vec{e_n}$ being essentially free of $1/f$ noise is determined by passive components (25) and may be superior to J-FET's at subaudio frequencies. However, for practically all applications this technique is now regarded obsolete.

In junction transistors, the low-frequency noise corner is current dependent due to m and the corner frequency is usually in the order of several hundred hertz. From expression (15), the excess noise in junction transistors is equivalent to a decrease in β. Consequently, the optimal current for a given source resistance would be lower at the flicker noise region than it would be at medium frequencies. It has also been found that $1/f$ noise can significantly increase as a result of avalanching the base-emitter junction [26]. This may happen during supply turn-on or input overloading and may be prevented by parallel protection diodes. Excess noise in junction transistors is observed also in $\vec{e_n}$ and is due to base-current shot-noise passing through $r_{b'}$. However, the corner frequency of $\vec{e_n}$ may be very low and is usually of no concern.

XIII. Impedance, Noise, and Negative Feedback

A preamplifier for a specific transducer should in general meet two main requirements: 1) the output voltage be proportional to the desired signal over the bandwidth of interest, and 2) the equivalent input noise should not exceed a certain minimum related to the signal expected amplitude. The first requirement is usually easier to comply with since one can always design a network which will restore a frequency distorted signal. To minimize the complexity of such a network and, perhaps, eliminate it altogether one should distinguish between two main types of transducers, as in Section V, i.e.: 1) those in which the desired signal is proportional to the source short-circuit current, such as reverse biased photodiodes or magnetic inductive transducers, and 2) those in which the signal is proportional to the open-circuit voltage, such as piezoelectric detectors. The first-order equivalent circuits of these two types of sources would be a current source with a shunt admittance, and a voltage source with a series impedance, respectively. For the current signal source, in order to eliminate the effect of the shunt admittance the ideal preamplifier would have zero input impedance, with an output voltage proportional to the input current. Similarly, for a voltage-type signal source, a high input impedance voltage amplifier is in order. Evidently, there are sources for which the equivalent circuit is more complex, necessitating a signal restoring network. One example is a reactive source with a noise tuning reactance (see Section VII). Another example is a flux measuring inductive sensor with internal resistance as discussed below.

In establishing an appropriate input impedance there is no need to compromise noise performance since noise is not

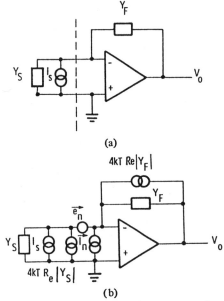

Fig. 15. (a) Parallel feedback amplifier with a current signal source at the input. (b) Total noise and signal at the input of the amplifier.

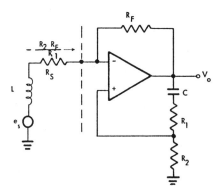

Fig. 16. Extending low-frequency response for inductive source with a negative input impedance amplifier.

necessarily associated with it. The reason is that using a negative feedback, one can modify the input impedance without significantly affecting the input equivalent noise sources. In contrast to this flexibility, the noise performance depends heavily on the input stage.

Fig. 15(a) shows a current signal source I_s, with internal admittance Y_s, terminated with a voltage amplifier with input admittance Y_{in}. In Fig. 15(b) the input network has been replaced by its Norton equivalent where the total input current noise is given by

$$\overline{I_n^2} + \overline{e_n^2} \cdot |Y_s|^2 + 4kTR_e[Y_s]. \quad (17)$$

Now, if we add a parallel feedback admittance Y_f, it will add to the source admittance since, for this purpose, the output side of the feedback admittance is at ground. Consequently, the total current noise at the input would be

$$\overline{I_n^2} + \overline{e_n^2}|Y_s + Y_f|^2 + 4kTR_e[Y_s + Y_f]. \quad (18)$$

The input network can be regarded as an equivalent current source comprised of a signal component and a noise component. When feedback is applied it obviously exerts the same effect on the signal source as on the total noise source so that their ratio remains unchanged at any frequency. However, this may well cause a change of the transfer function from input current to output voltage. So that if the spectral contents of the signal and noise are different, the integrated S/N over the bandwidth of interest may change. However, for both cases, as long as the transfer function is equalized, there would be no difference compared to the open loop. Matters become different, if for the sake of fair comparison we assume $Y_F = 0$ in the nonfeedback amplifier, which then theoretically appears to be superior. Practically, this means that depending on the feedback type, series or parallel, the impedance level of the feedback elements should be sufficiently small or sufficiently large to minimize the added noise.

The transimpedance amplifier in which $1/Y_F = R_F$ [27] provides an example where negative feedback serves for obtaining wide bandwidth by reducing the input impedance. It is advantageous for capacitive current sources since ideally no signal integration would take place at the input. Input impedance can certainly be made arbitrarily low with a parallel input resistor R; however, the thermal current noise $4kT/R$ would adversely affect the S/N. The transimpedance amplifier, on the other hand, has an input impedance which is equal to R_F/A, but the thermal noise added is that of the resistor R_F only. Thus a low noise and a large bandwidth can be achieved simultaneously if R_F and A are sufficiently large. Similarly a charge amplifier is obtained when $Y_F = \omega C$, i.e., the feedback

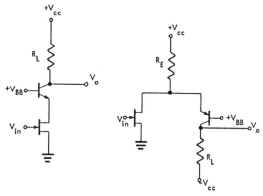

Fig. 17. Two configurations of the cascode connection.

is by means of capacitor (in parallel with a large bias resistor) and is ideal for sources represented by a voltage source in series with a capacitance.

Magnetic flux-measuring inductive signal sources such as magnetic reproducing heads and current transformers are another example where the signal is proportional to the short-circuit current. In practice, however, there is always certain winding resistance which would, in combination with the inductance, limit the low-frequency response even if the amplifier has zero input impedance. Neglecting parasitic capacitance, the source can be represented as in Fig. 16 where it is coupled to an amplifier with a combined positive and negative feedback. A negative input resistance is thus realized which can nearly cancel the winding resistance effect and appreciably extend the low-frequency cutoff. The thermal noise of the source resistance cannot, obviously, be eliminated. The feedback resistor R_F should be large enough to add a negligible thermal noise, while the positive feedback resistors R_1, R_2 should similarly be small enough. In the ideal case where the input resistance exactly cancels the source resistance the amplifier is in fact operated open loop. To ensure dc stability, the positive feedback is ac coupled. This circuit is simple and performs better than a high-input impedance amplifier and equalization network [28].

XIV. THE JUNCTION TRANSISTOR AND THE FET AS A FIRST STAGE

As shown in Section VI, the low-noise figure of merit of any amplifier or input device is the product of its two equivalent input noise sources at the frequency range of interest; theoretically at least, one should select a device in which $\overline{e_n^2} \cdot \overline{I_n^2}$ is at minimum, and the source impedance would then only determine the turns ratio of the noise matching transformer or the number of input devices to be connected in parallel.

On the basis of formulas (12), (12a), (13), and assuming that $r_{b'}$ is negligible, then neglecting the dependence of f_T and β on I_e, the figure of merit for the junction transistor would be

$$(2KT)^2 \left[\frac{1}{\sqrt{\beta}} + \frac{f^2}{f_T^2} \right].$$

The figure of merit for the FET, substituting $g_m/C_{gs} = f_T$ is found to be quite similar

$$(2KT)^2 \left[\frac{1.5 q I_g}{kT g_m} + \frac{f^2}{f_T^2} \right].$$

Thus, at low frequencies, high β and low I_g characterize low-noise devices, whereas at high frequencies, high cutoff is the figure of merit. It should be kept in mind that the FET would usually consume more power due to the dependence of g_m on I_D and that noise matching is the basic condition for the above comparison. Thus applying the above result to a relatively low impedance source may necessitate the use of a transformer with an impractical turns ratio or, alternatively, a large number of FET's in parallel. However, as already shown, exact noise matching is not critical.

In general, the FET is a better choice for high-impedance wide-band signal sources. Since, at low frequencies, high β junction transistors operated at sufficiently small collector currents may achieve base currents comparable to and at high temperatures even smaller than typical gate leakage currents. In FET's, on the other hand, the low noise performance spans a wider range of signal sources and frequencies. For very low impedance sources the junction transistor is superior due to its potentially higher g_m when accompanied with low $r_{b'}$. As far as noise is concerned the actual selection should eventually be made by a quantitative comparison of the integrated noise over the signal bandwidth.

As already mentioned, for both devices, the input noise sources are essentially independent on the configuration; however, the voltage gain, current gain, and input impedance are different for the common-base and the common-emitter. For example, due to its unity current gain, the common-base is a bad choice for a current-signal source. With voltage-signal sources, on the other hand, sufficient voltage gain may be obtained as long as R_s is of the order of $1/g_m$. As a result the common-base can be preferable when its low input impedance is an advantage, mainly for source impedance matching in communications applications and in general, where its wide bandwidth and low reverse-capacitance are important as in infrared low resistance with bandwidth HgCdTe detectors.

In most cases, the common-emitter or common-source are preferable as a first stage owing to the high gain and input impedance. However, the voltage gain A of such stage tends to increase the input capacitance due to Miller-effect by $A C_r$ where C_r is the reverse capacitance of the device. (C_r tends to be very small in D-type MOSFET's [26] and some junction FET's–see Section X). In other than low-frequency amplifiers this is undesirable, and a classical input configuration, the cascode, is often used to minimize this effect (see Fig. 17). In this configuration the second stage which serves as a load for the first stage is a common-base operated at sufficient current so as to decrease the voltage gain of the input stage, typically not much greater than unity. The common base is a unity current amplifier and the total voltage gain is $g_m R_L$. Thus the cascode is a combination of unity voltage and unity current gain stages and has high speed capabilities, besides low input capacitance. The noise of the second stage can be accounted for by vectorially adding its base-current noise, divided by g_{m1}, to the first stage noise-voltage source. As already mentioned this base noise tends to increase at low as well as at high frequencies. Thus a junction FET second stage may sometimes be preferable.

Some care is usually necessary in the selection of passive components in low noise design [10, ch. 9]. As already mentioned, carbon composition resistors in particular, as well as various potentiometers develop extra noise, which is proportional to the dc current. Usually, metal film resistors are the best for an input stage. In addition, electrolytic capacitors may contribute noise due to leakage current and should be avoided if possible.

Acknowledgment

The author wishes to thank Dr. Neal Butler for his helpful comments and suggestions.

References

[1] M. E. Gruchalla, "Measure wide-band white noise using a standard oscilloscope," *Electron Devices Newsletters*, June 5, 1980.
[2] W. R. Davenport and W. L. Root, *An Introduction to the Theory of Random Signals and Noise*. New York: McGraw-Hill, 1958.
[3] V. Radeka, "1/f noise in physical measurements," *IEEE Trans. Nuclear Sci.*, vol. NS-16, pp. 17-35, Oct. 1969.
[4] R. A. Dukelow, "An experimental investigation of very low frequency semiconductor noise," Ph.D. dissertation, California Inst. Technol., Pasadena, 1974.
[5] E. M. Cherry and D. E. Hooper, *Amplifying Devices and Low-Pass Amplifier Design*. New York: Wiley, 1968, p. 39.
[6] H. A. Haus et al., "Representation of noise in linear twoports," *Proc. IRE*, vol. 48, pp. 69-78, Jan. 1960.
[7] R. W. A. Ayre, "A new transistor noise test set," *Proc. IEEE*, vol. 60, p. 151, 1972.
[8] Y. Netzer, "A new interpretation of noise reduction by matching," *Proc. IEEE*, vol. 62, Jan. 1974.
[9] —, "Low noise optimization of J-FET input stage for capacitive current sources," *Proc. IEEE*, vol. 65, July 1977.
[10] C. D. Motchenbacher and F. C. Fitchen, *Low Noise Electronic Design*. New York: Wiley, 1973.
[11] E. Faulkner and D. W. Harding, "Some measurements on low noise transistors for audio frequency applications," *Radio Electronic Eng.*, July 1968.
[12] "Matching of a whip aerial to a transistorized V.H.F. receiver," *Electron. Appl.*, vol. 24, no. 1.
[13] T. E. Wade et al., "Noise effects in bipolar junction transistors at cryogenic temperatures: Parts I, II," *IEEE Trans. Electron Devices*, vol. ED-23, Sept. 1976.
[14] F. M. Klassen, "High frequency noise of the Junction Field Effect Transistor," *IEEE Trans. Electron Devices*, vol. ED-14, pp. 368-373, July 1967.
[15] M. B. Das, "FET noise sources and their effects on amplifier performance at low frequencies," *IEEE Trans. Electron Devices*, vol. ED-19, Mar. 1972.
[16] P. D. LeVan, "Preamplifier noise in indium-antimonide detector systems," *SPIE Proc.*, vol. 242, 1980.
[17] S. S. Sesnic and G. R. Craig, "Thermal effects in JFET and MOSFET devices at cryogenic temperatures," *IEEE Trans. Elect.*, ED-19, no. 8, Aug. 1972.
[18] R. A. Pease, "Low-noise composite amp beats monolithics," *Electron Devices Newsletters*, May 5, 1980.
[19] S. T. Hsu, "Noise in high gain transistors and its application to the measurement of certain transistor parameters," *IEEE Trans. Electron Devices*, vol. ED-18, July 1971.
[20] M. B. Das, "On the current dependence of low frequency noise in bipolar transistors," *IEEE Trans. Electron Devices*, vol. ED-22, Dec. 1975.
[21] R. C. Jaeger and A. J. Brodersen, "Low frequency noise sources in bipolar junction transistors," *IEEE Trans. Electron Devices*, vol. ED-17, Feb. 1970.
[22] T. Koji, "The effect of emitter-current density on popcorn noise in transistors," *IEEE Trans. Electron Devices*, vol. ED-22, Jan. 1975.
[23] M. Stoisiek and D. Wolf, "Origin of 1/f noise in bipolar transistors," *IEEE Trans. Electron Devices*, vol. ED-27, Sept. 1980.
[24] J. W. Hasliett and E. J. M. Kendall, "Temperature dependence of low-frequency excess noise in junction-gate FET's, *IEEE Trans. Electron Devices*, vol. ED-19, Aug. 1972.
[25] J. R. Biard, "Low-frequency reactance amplifier," *Proc. IEEE*, Feb. 1963.
[26] B. A. McDonald, "Avalanche-induced 1/f noise in bipolar transistors," *IEEE Trans. Electron Devices*, vol. ED-17, Feb. 1970.
[27] Y. Netzer, "Simplify fiber optic receivers with a high quality preamp," *Electron Devices Newsletters*, Sept. 20, 1980.
[28] —, "Negative resistance amplifier improves current probes low frequency performance," *Electron Devices Newsletters*, Nov. 5, 1980.

Comments on "The Design of Low-Noise Amplifiers"

RICHARD G. MARTIN

In the above titled paper,[1] equation (10) is incorrect. The equation should read:

$$NF = 10 \log_{10} \left[\frac{(S/N)_{\text{in}}}{(S/N)_{\text{out}}} \right].$$

Manuscript received June 17, 1981; revised July 15, 1981.
The author is with Hughes Aircraft Company, 1310 Mt. Hood Street, Las Vegas, NV 89110.
[1] Y. Netzer, *Proc. IEEE*, vol. 69, no. 6, p. 728, June 1981.

This is probably a typographical error, but I find this mistake common in papers on Noise Figure.

Reply[2] *by Y. Netzer*[3]

Concerning Mr. Martin' remark regarding my paper[1] on low-noise design, the above formula is indeed incorrect. There also are two additional errors.

1) In the first formula on page 740, β should appear without the square root sign.
2) The sources cited in references [1], [18], [27], [28] should read EDN.

[2] Manuscript received July 15, 1981.
[3] Y. Netzer is with Honeywell EOO, Lexington, MA 02173.

Amplifier Techniques for Combining Low Noise, Precision, and High-Speed Performance

GEORGE ERDI, SENIOR MEMBER, IEEE

Abstract—A monolithic operational amplifier is presented which optimizes voltage noise both in the audio frequency band, and in the low frequency instrumentation range. In addition, the design demonstrates that the requirements for low noise do not necessitate compromising the specifications in other respects. Techniques are set forth for combining low noise with high-speed and precision performance for the first time in a monolithic amplifier.

Achieved results are: 3 nV/$\sqrt{\text{Hz}}$ white noise, 80 nV$_{p-p}$ noise from 0.1 to 10 Hz, 17 V/μs slew rate, 63 MHz gain-bandwidth product, 10 μV offset voltage, 0.2 μV/°C drift with temperature, 0.2 μV/month drift with time, and a voltage gain of two million.

I. Introduction

THE constantly improving designs of analog circuits have reduced the error contribution of most parameters to the "noise level." In many applications, noise does become the limiting factor on performance. In the case of operational amplifiers, the error due to a specific parameter can always be controlled: temperature drift effects can be reduced by regulating the environment of the system, gain error terms can be minimized by cascading several amplifying stages, etc. Voltage noise, however, cannot be eliminated, and, therefore, it can be defined as the ultimate error source.

The major emphasis of the monolithic operational amplifier design presented here was to minimize voltage noise both in the audio frequency range and in the low frequency instrumentation range. Details of this effort are described in Section II. The precision characteristics of the design are discussed in Section III. Several examples illustrate how low noise and precision can be complementary requirements.

Section IV considers the high-speed aspects of the design; the combination of low noise and bandwidth broadening is discussed. Achieved performance is summarized in Section V.

II. Low Noise Design

The noise spectrum of a typical operational amplifier is shown in Fig. 1. In the audio region, noise is flat or white noise, and is characterized by a constant value over all frequencies of interest. The low frequency instrumentation range noise is usually called the 1/f region because

$$(\text{voltage noise})^2 \propto 1/f. \qquad (1)$$

Fig. 1 depicts voltage noise, but current noise has the same form, i.e., it is completely characterized by its white noise value and the location of its 1/f corner frequency. The amplifier's noise contribution in a band from frequencies f_1 to f_2, ($N_{f_2-f_1}$), can be determined [1] from

$$N_{f_2-f_1} = N_0 \left[f_0 \ln \frac{f_2}{f_1} + (f_2 - f_1) \right]^{1/2} \qquad (2)$$

where

$N_{f_2-f_1}$ can be either voltage or current noise,
N_0 is the white voltage or current noise density (usually specified in nV/$\sqrt{\text{Hz}}$ or pA/$\sqrt{\text{Hz}}$), and
f_0 is the corner frequency where the 1/f and white noise components intersect.

In the audio range noise is minimized by a simple reduction of white noise; in the instrumentation region the problem is more complicated. In addition to low white noise, the 1/f corner frequency f_0 has to be as low as possible. Reducing white noise and f_0 are two separate tasks; low white noise does not necessarily imply low instrumentation range noise, as illustrated in Fig. 2.

Here the noise spectra of three operational amplifiers are shown. The popular 741 has relatively high white noise and f_0; it cannot be classified as a low noise amplifier in any region. The audio op amp [2] has low white noise, but because its 1/f corner is high at 70 Hz, its low frequency noise is rather high. The amplifier being described here (type OP-27/37) has minimum white noise and a low f_0 of 2.7 Hz.

A. White Noise Reduction

The essential requirement for low noise is to minimize the number of components, transistors, or resistors, contributing to input noise. The voltage noise of the simple, resistively loaded differential input stage of Fig. 3 depends on input transistors $Q1$ and $Q2$ only, provided that the noise of the load resistors R_L, and the input referred noise of the second stage are negligible.

The white voltage noise (e_{N_0}) of a differential pair [3] is given by

$$e_{N_0}^2 = 8kT \left(\frac{kT}{2qI_c} + r_{bi} + r_{be} \right) \qquad (3)$$

where

I_c is the collector current of the transistors,
r_{bi} is the intrinsic base resistance underneath the emitter, and

Manuscript received April 14, 1981; revised June 15, 1981.
The author was with Precision Monolithics, Inc., Santa Clara, CA 95050. He is now with the Linear Technology Corporation, Mountainview, CA 94043.

Reprinted from *IEEE J. Solid-State Circuits*, vol. SC-16, no. 6, pp. 653-661, Dec. 1981.

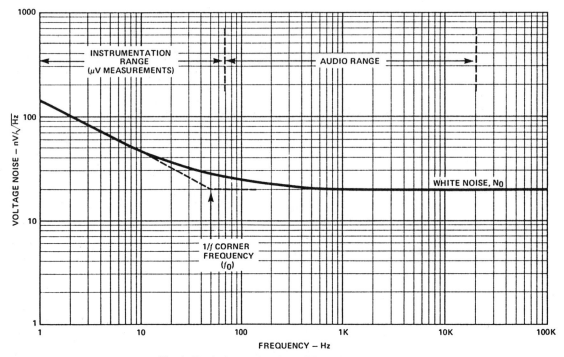

Fig. 1. Typical operational amplifier noise spectrum.

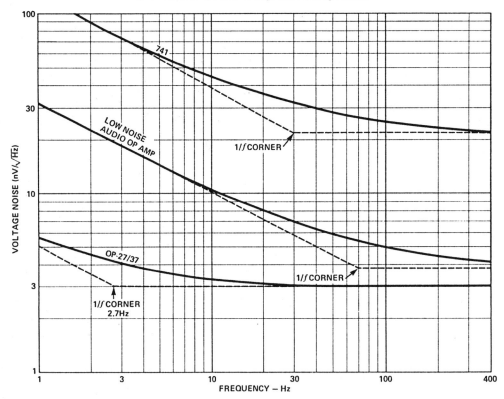

Fig. 2. Noise spectra of three operational amplifiers.

r_{be} is the extrinsic base resistance from the base contact to the edge of the emitter; it also includes base and emitter interconnection and contact resistances.

The design goal of the low noise op amp was to achieve a white noise of 3 nV/$\sqrt{\text{Hz}}$. This implies that the total resistance of the terms in parentheses of (3) is 260 Ω. The collector current dependent first term contributes 110 Ω by operating the input stage at 120 μA; r_{bi} and r_{be} are minimized by long and narrow input transistor emitters surrounded by base contacts.

Earlier, it was assumed that the second stage and load resistor noise contributions are, or at least can be made, negligible. The voltage gain A_{Vi} of the input stage is

$$A_{Vi} = g_m R_L = \frac{qI_c}{kT} R_L = \frac{120\,\mu A}{26\,mV} \times 22\,k\Omega \simeq 100. \quad (4)$$

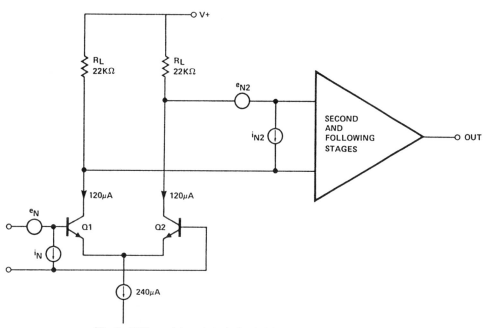

Fig. 3. Differential, resistively-loaded input stage with input and second-stage noise sources.

If the second stage voltage noise is less than 50 nV/$\sqrt{\text{Hz}}$, which is not a stringent condition, its input referred contribution will be less than 0.5 nV/$\sqrt{\text{Hz}}$, which increases input voltage noise by less than 1.4 percent when root sum squared with 3 nV/$\sqrt{\text{Hz}}$. Similarly, the load resistor noise [=(2 × $4kTR_L$)$^{1/2}$] referred to the input will be a negligible 0.27 nV/$\sqrt{\text{Hz}}$.

A general relationship can be developed between the input transistor noise (e_{N_0}) and the input referred resistor noise (N_R/A_{vi}) of the differential stage. From (3),

$$e_{N_0}^2 > \frac{4(kT)^2}{qI_c}. \tag{5}$$

Also,

$$\left(\frac{N_R}{A_{vi}}\right)^2 = \frac{2 \times 4kTR_L}{(g_m R_L)^2} = \frac{8(kT)^2}{qI_c A_{vi}}. \tag{6}$$

Therefore,

$$\frac{e_{N_0}}{N_R/A_{vi}} > \left(\frac{A_{vi}}{2}\right)^{1/2}. \tag{7}$$

From (7), if the differential gain is large, the resistor noise will be negligible. For example, if the gain is greater than 20, the resistor noise contribution will be less than 5 percent.

The current noise i_{N2} of the second stage flows through the load resistors, and thus creates an input referred voltage noise component $e_N(i_{N2})$:

$$e_N(i_{N2}) = 2i_{N2} R_L/A_{vi} = \frac{2i_{N2}}{g_m}. \tag{8}$$

The transimpedance of the input state is only 220 Ω. Therefore, as long as the white current noise of the second stage is less than 3 pA/Hz$^{1/2}$, its influence will be negligible. This again is not a stringent condition.

Fig. 4 illustrates three commonly used operational amplifier input stages. Fig. 4(a) is the 741 input stage [4] and Fig. 4(b) is used as an input on many precision amplifiers [5]. Both of these have active loads ($Qa5$, $Qa6$, $Qb3$, $Qb4$). Active loads, by their very nature, amplify their own internal noise. This current noise then flows through the input transistors, thereby degrading noise performance. A detailed noise analysis of active load stages can be found in [6].

Fig. 4(c) shows a resistively loaded input stage which is employed on the most popular three-gain-stage op amps [7], [8]. With its collector current at 8 μA, the voltage noise at 9 nV/(Hz)$^{1/2}$ is basically limited by the transistor noise of $Qc1$ and $Qc2$ and can be calculated from (3). When the collector current is increased to 120 μA, and the input device geometries are redesigned to minimize r_{bi} and r_{be}, the observed noise of Fig. 4(c) will be higher than predicted by (3). Secondary noise contributors which are negligible at the 9 nV/(Hz)$^{1/2}$ level suddenly become significant noise sources. For example, the input bias current cancellation scheme ($Qc3$-$Qc10$) adds about 2 nV/(Hz)$^{1/2}$ to the total voltage noise. Resistors $R1$ and $R2$ limit the current through the input protection diodes when large differential voltages are applied. However, the 1 kΩ total source resistance contributes 4.1 nV/(Hz)$^{1/2}$ of noise.

On the present design, the limiting resistors are eliminated, and the bias current cancellation network is removed from the signal path—to be discussed later—resulting in the simple input stage of Fig. 3.

B. Current Noise

Voltage noise is inversely proportional to the square root of collector current as shown by (3). Current noise, however, is directly proportional to the same function. Therefore, an inevitable byproduct of reduced voltage noise is increased current noise. The amplifier's current noise is shown in Fig. 5; it has the form described by (2), and the critical parameters

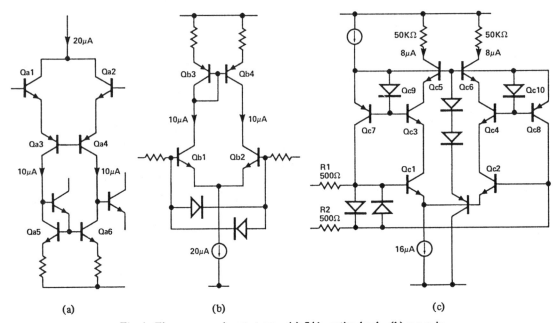

Fig. 4. Three op amp input stages. (a) 741–active load. (b) n-p-n input–active load. (c) Resistive load with bias current cancellation.

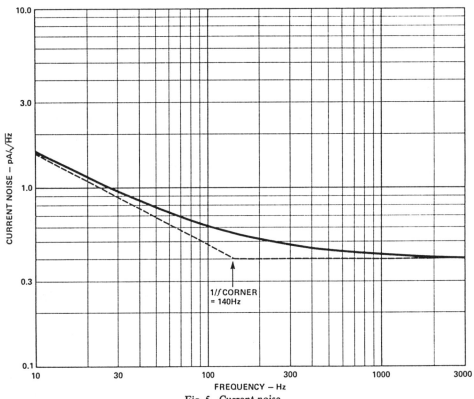

Fig. 5. Current noise.

again are the magnitude of the white noise and the location of the $1/f$ corner frequency.

The total observed noise of an op amp is

$$(\text{total noise})^2 = (\text{voltage noise})^2 + (\text{source resistor noise})^2$$
$$+ (\text{current noise} \times \text{source resistor})^2. \quad (9)$$

The obvious contribution of current noise is the last term of (9). In addition, voltage noise will have current noise dependent components, e.g., the current noise flowing through r_{bi} and r_{be} of the input transistors. Furthermore, internal current noise sources create voltage noise, as shown earlier in (8).

White current noise is seldom a problem when total noise is considered. Although white noise is relatively high, it still does not limit performance. Fig. 6 plots total noise as a function of source resistance. 1000 Hz noise is dominated by resistor noise when source resistance is in excess of 1 kΩ.

The key parameter in considering current noise is the location of the $1/f$ corner frequency. A survey of op amp data sheets indicates corner frequencies of 200 Hz–2 kHz, typi-

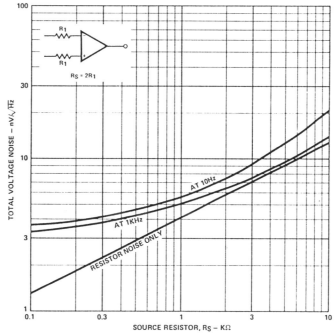

Fig. 6. Total voltage noise versus source resistance; total noise = $[e_N^2 + (i_N R_S)^2 + 4kTR_S]^{1/2}$.

cally an order of magnitude higher than the f_0 of voltage noise. The current noise f_0 is strictly process dependent, it can be as high as 10 MHz for some digital processes [6]. For this device, the $1/f$ corner is low at 140 Hz (Fig. 5), partially due to the silicon nitride passivation used, which acts as an additional gettering step.

When low frequency (10 Hz) total noise is plotted in Fig. 6, the resistor noise is unchanged, but current noise is enlarged four fold. Voltage noise is only slightly increased (Fig. 2). With source resistors in excess of 5 kΩ, current noise starts to dominate. The total noise at this point, however, is four times higher than voltage noise. The optimized voltage noise of the device is completely wasted, not because of current noise, but because source resistor noise exceeds voltage noise even at 1 kΩ.

C. Minimizing Low Frequency Noise

The $1/f$ region of voltage noise is basically a current noise-caused phenomenon. It occurs because current noise at some internal node in the circuit flows through a relatively large resistor, creating voltage noise. As shown earlier, in the white current noise region, the problem is insignificant. However, because of the relatively high $1/f$ corner of current noise, at low frequencies the current noise-caused voltage noise can easily increase by an order of magnitude.

A reduction of instrumentation range voltage noise therefore requires a low $1/f$-corner current noise process (which has been achieved here), and a systematic evaluation of the internal nodes of the circuit. Buffering with emitter followers should be used when either the current noise or the impedance is high at a given node.

An example of this is shown in the simplified schematic of the operational amplifier (Fig. 7). With an n-p-n input stage a lateral p-n-p second stage is always necessary for level shifting.

In many designs the bases of the lateral p-n-p transistors ($Q23$, $Q24$) are tied directly to the input stage. However, the low frequency current noise of these p-n-p's is very high for several reasons. They operate at high emitter currents (240 μA) where their current gains have already fallen off to a low value. In addition, the lateral p-n-p being a surface device, its $1/f$ corner frequency is significantly higher than that of the n-p-n transistors. In this particular example, the white current noise of $Q23$, $Q24$ is 2.5 pA/Hz$^{1/2}$, its $1/f$ corner occurs at 500 Hz. Converting this second stage current noise to input referred voltage noise using (8), gives 3 nV/$\sqrt{Hz}$ at about 70 Hz. In other words, loading the input stage directly with lateral p-n-p's $Q23$, $Q24$ would move the $1/f$ voltage noise corner of the amplifier from 2.7 to 70 Hz.

The insertion of emitter followers $Q21$ and $Q22$ completely eliminates this problem. The current noise of $Q21$ and $Q22$ is actually $\sqrt{2}$ times less than the input current noise because of the lower operating current. The current noise of $Q23$ and $Q24$ flows through two 1 kΩ resistors (the output impedance of the emitter followers and the base resistance of the lateral p-n-p's) rather than the 22 kΩ load resistors. The input referred voltage noise contribution of the $Q23$, $Q24$ current noise at 2.7 Hz is only 0.7 nV/Hz$^{1/2}$, while the $Q21$, $Q22$ current noise translates to 0.9 nV/Hz$^{1/2}$.

The $1/f$ corner is at 2.7 Hz because at that frequency the root sum squared of all the frequency dependent noise sources equals the white noise of 3 nV/Hz$^{1/2}$. The dominant terms are the input current noise (which is 3.1 pA/Hz$^{1/2}$ at 2.7 Hz) flowing through the equivalent input resistance of 520 Ω, as given by (3), and the input referred voltage noise of the second stage. The second stage noise is about 160 nV/Hz$^{1/2}$ at 2.7 Hz, or twice the noise of a 741 amplifier (80 nV/Hz$^{1/2}$ at 2.7 Hz as shown in Fig. 2).

The 0.1–10 Hz peak to peak noise of the op amp is 80 nV,

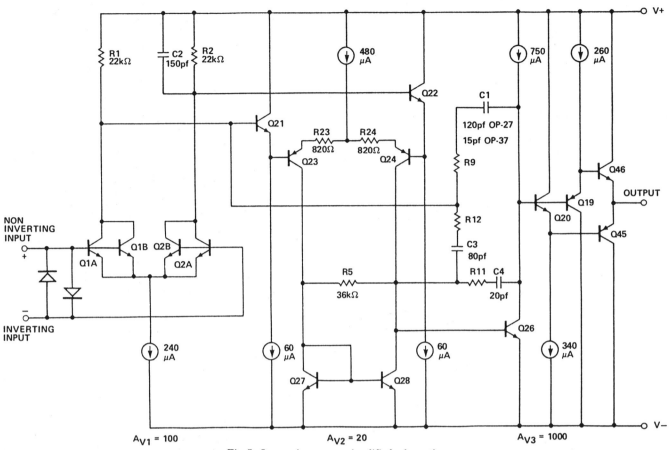

Fig. 7. Low noise op amp simplified schematic.

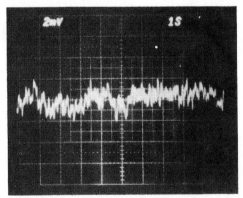

Fig. 8. Low frequency noise: 0.1-10 Hz peak to peak. 40 nV/div referred to input (closed-loop gain = 50 000).

as illustrated by the oscilloscope photograph of Fig. 8. The rms noise in this frequency band is 14.2 nV as calculated from (2).

III. Precision Design

The circuit employs all the well-established design techniques for achieving precision performance. The simple resistively loaded input stage has been demonstrated in the past to be the best for low offset voltage and drift with time and temperature [9]. In addition, the load resistors are ideal for on wafer Zener-zap adjustment of offset voltage to a few microvolts [8], which is also used on this circuit. And, as demonstrated in the previous section, resistive loading optimizes noise.

The quad-connection of input transistors [9] is another design tool which enhances both precision and low noise performance. The transistors making up the differential input pair are formed from cross connected segments of a quad of transistors ($Q1A$, $Q1B$, $Q2A$, $Q2B$ of Fig. 7). This has the well-known benefits of cancelling thermal gradients and variations in the epi and diffusions. As far as noise is concerned, the quad connection also helps because it effectively halves r_{bi} and r_{be} by the use of transistors in parallel.

Because the input stage operates at a collector current which is an order of magnitude higher than the typical 10 μA of most op amps, input bias current (I_B) can be a significant error contributor. Super β input transistors cannot be used to reduce I_B, because the intrinsic resistance of super β transistors underneath the emitter [r_{bi} of (1)] is inherently high. Therefore, the noise performance of a super β transistor is considerably worse than that of the equivalent n-p-n transistor (i.e., same device geometry and operating at the same current).

The bias current cancellation circuit of Fig. 9 provides the best compromise. As implied in Section II, it is removed from the signal path and therefore does not contribute to voltage noise or to input offset voltage. $Q11$ and $Q12$ are identical to the input transistors, and operate at the same current density and approximately the same collector-base voltage as $Q1$ and $Q2$. Therefore, the base current of $Q11$, $Q12$ precisely matches

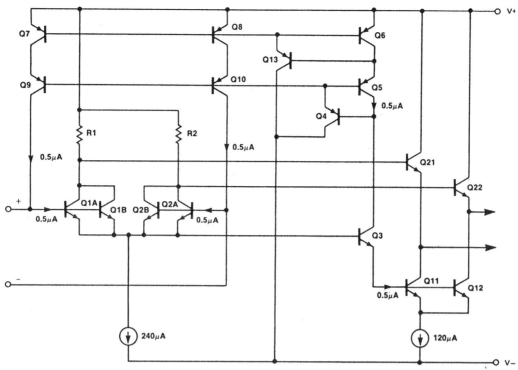

Fig. 9. Input bias current cancellation circuit.

the uncompensated input current of $Q1$ and $Q2$. The output of the precision current mirror of $Q5$-$Q10$ is fed back to the bases of the input transistors, cancelling the base currents. This scheme is successful in removing 98 percent of the input bias current and has a 3 GΩ common-mode input resistance.

Bias current compensation schemes, as a rule, increase input current noise by a factor of $\sqrt{2}$ because the cancelling current noise is uncorrelated to the input transistor current noise. This is the case of the bias current cancellation network used on the OP-07 amplifier [8] [Fig. 4(c)], where the noise currents of $Qc1$ and $Qc3$ are uncorrelated. In the circuit of Fig. 9, however, the noise currents of both $Q9$ and $Q10$ originate from the same source: the base current of $Q11$ and $Q12$, and thus correlate [10]. With balanced source resistors the cancellation noise currents represent a common-mode component and, therefore, do not add to the input current noise.

A necessary condition of precision performance is high voltage gain, preferably in excess of a million. This gain should be maintained even under heavy load conditions. Both thermal and electrical effects can prevent the realization of such high gain under load. The double-buffered output stage, shown on the simplified schematic of Fig. 7, isolates the load by a β^2 factor from the high impedance (approximately 80 kΩ) gain node at the collector of $Q26$. For positive swings the current gains of $Q46$ and $Q19$ are multiplied, for negative swings the β product of $Q45$ and $Q20$ applies. This β^2 multiplier is at least 5000, even when 10 mA is delivered to a 1 kΩ load, i.e., the reflected impedance at the collector of $Q26$ is more than 5 MΩ, or the electrical gain degradation is less than 2 percent.

Thermal feedback—the effect of output power dissipation

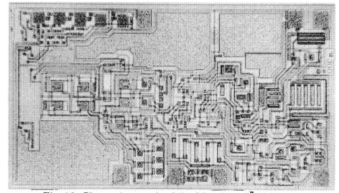

Fig. 10. Photomicrograph of the 96 × 54 mil^2 chip.

changes on the input transistors—is less than 1 μV. This is accomplished by a thermally symmetrical layout [9], which, by now, is a common technique of all precision amplifier designs, and can be observed on the chip photograph of Fig. 10. Fig. 11 shows the voltage gain with 1 kΩ load as measured on a Tektronix 178 tester. The straightness of this line illustrates the absence of thermal feedback, even as 10 mA is delivered to the load.

IV. HIGH-SPEED DESIGN

The relatively high operating current of the input stage (120 μA), which is necessary for low voltage noise, also provides an opportunity for increasing bandwidth. The typical three stage op amp's first stage gain is limited to 20 because of its 10 μA collector currents, and maximum practical load resistors of 50 kΩ. Here the input differential gain is 100. This "excess" gain allows the design of a wider band, less

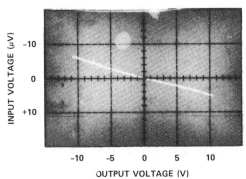

Fig. 11. Voltage gain, $R_L = 1$ kΩ (measured on Tektronix 178 linear tester).

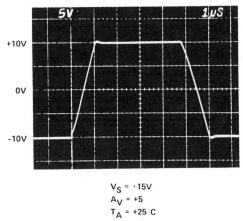

Fig. 12. Large signal transient response.

TABLE I
TYPICAL AND GUARANTEED SPECIFICATIONS OF THE LOW NOISE, PRECISION, HIGH-SPEED OP AMP AT $V_s = \pm 15$ V, $T_A = 25°$C

	Typ	Min/Max	Units
Noise Specifications			
Voltage noise: 0.1–10 Hz	80	180	nV$_{p-p}$
$f_0 = 10$ Hz	3.5	5.5	nV/$\sqrt{Hz}$
$f_0 = 1$ kHz	3.0	3.8	nV/$\sqrt{Hz}$
Current noise: $f_0 = 10$ Hz	1.6	4.0	pA/$\sqrt{Hz}$
$f_0 = 1$ kHz	0.4	0.6	pA/$\sqrt{Hz}$
Precision Specifications			
Offset voltage	10	25	μV
drift with temperature	0.2	0.6	μV/°C
drift with time	0.2	1.0	μV/mo
Input bias current	10	40	nA
Input offset current	7	35	nA
Voltage gain	2000	1000	V/mV
CMRR	126	114	dB
Speed Specifications			
Slew rate, $A_{VCL} \geq 1$ (OP-27)	2.8	1.7	V/μs
Slew rate, $A_{VCL} \geq 5$ (OP-37)	17	11	V/μs
Gain at 10 kHz (OP-37)	6.3	4.5	V/mV
Unity gain bandwidth (OP-27)	8.0	5.0	MHz
Other Specifications			
Power consumption	90	140	mW
Output voltage swing, $R_L \geq 600$ Ω	11.5	10	V
Capacitive load capability	2000	–	pF

accurate second stage without adversely influencing input accuracy. The usually slow lateral p-n-p level-shift amplifier is broadbanded with degenerating resistors $R23$ and $R24$ (Fig. 7) [11] and a low, controlled gain of 20. The frequency characteristics of the second stage are mainly determined by $R5$, $R23$, and $R24$, and to a reduced extent, by the lateral p-n-p's, $Q23$ and $Q24$. As a result, feed-forward capacitor $C3$ can bypass the second stage at a significantly higher frequency than in previous three stage precision op amp designs.

Capacitor $C1$ sets the dominant pole. $C2$ makes the high frequency signal single-ended, i.e., it rolls off the gain of the input stage on the side which is not fed forward. The use of resistors $R9$, $R11$, and $R12$ allows shaping of the frequency response with appropriately placed zeros to cancel poles occurring in the 5–20 MHz range.

On the unity gain compensated version of the design (type OP-27) $C1$ is 120 pF, the bandwidth is 8 MHz with 70° phase margin. $C1$ is reduced to 15 pF on the decompensated model (type OP-37) which is stable in closed loop gains of five or more. On this device slew rate is 17 V/μs (Fig. 12), voltage gain at 10 kHz is still 6300.

Wider bandwidth has beneficial effects as far as precision performance is considered, specifically, gain error at low frequencies. A typical precision op amp may have a dc gain of two million, with a bandwidth of 600 kHz. Since the gain rolls off with frequency at a 20 dB/decade rate, the full voltage gain of the amplifier can only be realized at frequencies below the dominant pole of 0.3 Hz. Many low frequency instrumentation applications, of course, have to process signals which change at a faster rate. The OP-27's dominant pole occurs at 7 Hz, the OP-37's at 30 Hz.

V. PERFORMANCE

The achieved specifications of the circuit are listed in Table I. The significance of most parameters has already been discussed. The device has been in high volume production since December 1980; thus, the validity of the specifications has been demonstrated by more than just a few developmental units.

VI. CONCLUSIONS

The amplifier described advances the state of the art by a significant reduction of noise simultaneously with enhanced precision and high-speed performance.

ACKNOWLEDGMENT

The author wishes to recognize the contribution of T. S. Bernardi and Y. Gakhnokhi in optimizing the compensation network and in characterizing the device. Giok Bing Wu was responsible for mask design, and A. Honegger for test and product engineering. P. Scales typed the manuscript.

REFERENCES

[1] G. Erdi, "Noise performance of the Precision Monolithics SSS725 instrumentation operational amplifier," Precision Monolithics Application Note, 1972.
[2] TDA 1034 Data Sheet, Philips, Apr. 1976; NE5534 Data Sheet, Signetics, 1978.
[3] A. Willemsen and N. Bel, "Low base resistance integrated circuit transistor," *IEEE J. Solid-State Circuits*, vol. SC-15, p. 245, Apr. 1980.
[4] D. Fullagar, "A new high-performance monolithic operational amplifier," Fairchild Semiconductor Applications Brief, May 1968.

[5] M. A. Maidique, "A high precision monolithic super-beta operational amplifier," *IEEE J. Solid-State Circuits*, vol. SC-7, p. 480, Dec. 1972.

[6] P. R. Gray and R. G. Meyer, *Analysis and Design of Analog Integrated Circuits*. New York: Wiley, 1977, ch. 11.

[7] G. Erdi, "Instrumentation operational amplifier with low noise, drift, bias current," in *Northeast Res. Eng. Meeting Rec. Tech. Papers*, Oct. 1972; also OP-05 Data Sheet, Precision Monolithics, Inc., Jan. 1973.

[8] —, "A precision trim technique for monolithic analog circuits," *IEEE J. Solid-State Circuits*, vol. SC-10, p. 412, Dec. 1975; also OP-07 Data Sheet, Precision Monolithics, Inc., June 1974.

[9] —, "A low drift, low noise monolithic operational amplifier for low level signal processing," Fairchild Semiconductor, Application Brief 136, July 1969.

[10] Y. Nishikawa and J. E. Solomon, "A general purpose wideband operational amplifier," in *ISSCC Dig. Tech. Papers*, 1973, pp. 144-145.

[11] J. E. Solomon, "The monolithic op amp: A tutorial study," *IEEE J. Solid-State Circuits*, vol. SC-9, p. 322, Dec. 1974.

Low-Frequency Noise Considerations for MOS Amplifiers Design

JEAN-CLAUDE BERTAILS

Abstract—Equivalent input noise voltages of MOS amplifiers working in a low-frequency range have been calculated in the three basic technologies, i.e., single-channel enhancement-load, single-channel depletion-load and CMOS. Means of reducing that noise are discussed and practical results given for CMOS technology.

I. INTRODUCTION

It is well known that the noise of MOS amplifiers working in a low-frequency range is high, due to the dominant contribution of flicker noise, or $1/f$ noise. Many investigators have developed theories [1], [2] that generally explain flicker noise by a mechanism of current carrier trapping related to surface-states energically distributed within the semiconductor bandgap. Most of them establish that equivalent gate-noise voltage over a given frequency-band decreases with transistor-gate area and is only slightly dependent on the gate-bias voltage. According to them, the gate-noise voltage can be written

$$e_n = \sqrt{\frac{a_n}{ZL}} \qquad (1)$$

where Z and L are, respectively, the width and the length of the channel and a_n is a parameter depending on technology and, slightly, on bias. However, as it will be shown, the input noise is not reduced by increasing the ZL product of the input transistor.

Using the same simple models and notation as [3] recalled in the Appendix, it is possible to calculate the equivalent input noise voltages of the three usual amplifiers: enhancement-load, depletion-load, and CMOS amplifiers.

Single-channel amplifiers are given for NMOS technology, but results are straightforward extended to PMOS technology.

II. SINGLE-CHANNEL ENHANCEMENT-LOAD TECHNOLOGY

The stage is shown in Fig. 1. The input transistor $M1$ and the load transistor $M2$ are both saturated and the voltage gains for the input signal and for the noise voltages e_{n1} and e_{n2} are easily established [3].

For the input voltage and for e_{n1},

$$A_{V1} = -\frac{1}{1+\lambda 2} \sqrt{\frac{Z_1 L_2}{Z_2 L_1}}. \qquad (2)$$

For e_{n2},

$$A_{V2} = \frac{1}{1+\lambda 2}. \qquad (3)$$

Manuscript received November 11, 1978; revised January 25, 1979.
The author was with the Semiconductor Division SESCOSEM, Thomson-CSF, Saint-Egreve, France. He is now with EFCIS, Grenoble Cedex, France.

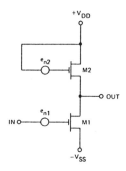

Fig. 1. Single-channel enhancement-load amplifier.

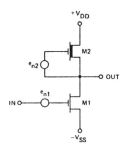

Fig. 2. Single-channel depletion-load amplifier.

Since the two noise voltages are uncorrelated, the equivalent input noise voltage is given by

$$e_n(\text{IN}) = \frac{1}{A_{V1}} \sqrt{(A_{V1} e_{n1})^2 + (A_{V2} e_{n2})^2} \qquad (4)$$

or

$$e_n(\text{IN}) = e_{n1} \sqrt{1 + \left(\frac{A_{V2} e_{n2}}{A_{V1} e_{n1}}\right)^2}. \qquad (5)$$

Using (1)–(3), we get

$$e_n(\text{IN}) = \sqrt{\frac{a_n}{Z_1 L_1} \left[1 + \left(\frac{L_1}{L_2}\right)^2\right]}. \qquad (6)$$

That formula shows that the equivalent input noise voltage (a) decreases when increasing the width Z_1 of the input transistor and the length L_2 of the load transistor, (b) has a minimum value versus the length L_1 of the input transistor, here for $L_1 = L_2$ and (c) is independent of the width Z_2 of the load transistor.

III. SINGLE-CHANNEL DEPLETION-LOAD TECHNOLOGY

The stage, shown in Fig. 2, includes an enhancement-mode input transistor $M1$ and a depletion-mode load transistor $M2$. The voltage gain for the input signal and the e_{n1} noise voltage is given by

$$A_{V1} = -\frac{1}{\lambda 2} \sqrt{\frac{Z_1 L_2}{Z_2 L_1}} \qquad (7)$$

Reprinted from *IEEE J. Solid-State Circuits*, vol. SC-14, no. 4, pp. 773–776, Aug. 1979.

and for e_{n2} noise voltage

$$A_{V2} = \frac{1}{\lambda 2}. \quad (8)$$

The equivalent input noise voltage is given by

$$e_n(\text{IN}) = \sqrt{\frac{a_{n1}}{Z_1 L_1}\left[1 + \frac{a_{n2}}{a_{n1}}\left(\frac{L_1}{L_2}\right)^2\right]} \quad (9)$$

where a_{n1} and a_{n2} parameters, generally different, apply respectively to enhancement input transistor and depletion load transistor. Like with enhancement-load technology, input noise reduction is obtained by increasing Z_1 and L_2, and $e_n(\text{IN})$ has a minimum value versus L_1 for

$$L_1 = \sqrt{\frac{a_{n1}}{a_{n2}}} L_2. \quad (10)$$

IV. CMOS Technology

The amplifier stage shown in Fig. 3 consists of a p-channel input transistor $M1$ and an n-channel load transistor $M2$ biased by a fixed voltage V_N. Naturally, the following calculations are easily transposed when using the n-channel transistor as input transistor and the p-channel transistor as load transistor.

To calculate the voltage gains for the input signal and for the two noise voltages e_{n1} and e_{n2}, it is no longer possible to suppose that the output impedances r_{ds1} and r_{ds2} of the two saturated transistors are infinite. The voltage gain for the input signal and for e_{n1} is given by

$$A_{V1} = -g_{m1}(r_{ds1} // r_{ds2}) \quad (11)$$

and for e_{n2},

$$A_{V2} = -g_{m2}(r_{ds1} // r_{ds2}), \quad (12)$$

with the two transconductances g_{m1} and g_{m2} given by

$$g_{m1} = 2\sqrt{k_1'\left(\frac{Z_1}{L_1}\right)I_D} \quad (13)$$

$$g_{m2} = 2\sqrt{k_2'\left(\frac{Z_2}{L_2}\right)I_D} \quad (14)$$

where I_D is the bias drain current of the stage. The equivalent input noise voltage is

$$e_n(\text{IN}) = \sqrt{\frac{a_{n1}}{Z_1 L_1}\left[1 + \frac{k_2' a_{n2}}{k_1' a_{n1}}\left(\frac{L_1}{L_2}\right)^2\right]}. \quad (15)$$

Many applications need differential amplifiers and a typical input stage is shown in Fig. 4.

The gain voltage for input signal and noise voltages e_{n1} and e_{n3} is

$$A_{V1} = \tfrac{1}{2}(g_{m1} + g_{m2})(r_{ds1} // r_{ds2}). \quad (16)$$

For e_{n2} and e_{n4}

$$A_{V2} = g_{m2}(r_{ds1} // r_{ds2}). \quad (17)$$

If $g_{m1} = g_{m3}$ and $g_{m2} = g_{m4}$, the equivalent input noise voltage is

$$e_n(\text{IN}) = \sqrt{2\frac{a_{n1}}{Z_1 L_1}\left[1 + \frac{k_2' a_{n2}}{k_1' a_{n1}}\left(\frac{L_1}{L_2}\right)^2\right]}, \quad (18)$$

i.e., identical to (15) multiplied by $\sqrt{2}$.

Thus, like with single-channel technologies, a noise improvement is obtained by increasing Z_1 and L_2 and the equivalent

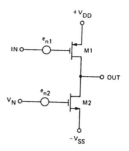

Fig. 3. CMOS amplifier.

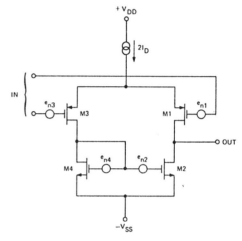

Fig. 4. CMOS differential amplifier.

noise input voltage has a minimum value versus L_1 for

$$L_1 = \sqrt{\frac{k_1' a_{n1}}{k_2' a_{n2}}} L_2. \quad (19)$$

It should be noted that for single-channel amplifiers, the L_1/L_2 ratio is generally taken low to obtain significant gain voltage, so that load transistor noise contribution in the equivalent input noise is low. But for CMOS amplifiers, this ratio does not contribute directly to the gain voltage of the stage, so that the noise due to the load transistor can be, without caution, very important. The next section example demonstrates a five times improvement of the noise voltage obtained by increasing the length of the load transistor.

Another amplifier stage, widely used with CMOS technology, is the push-pull stage shown in Fig. 5.

Three different voltage gains have to be considered. For the input signal

$$A_V = -(g_{m1} + g_{m2})(r_{ds1} // r_{ds2}). \quad (20)$$

For the e_{n1} noise voltage

$$A_{V1} = -g_{m1}(r_{ds1} // r_{ds2}). \quad (21)$$

And for the e_{n2} noise voltage

$$A_{V2} = -g_{m2}(r_{ds1} // r_{ds2}). \quad (22)$$

The equivalent input noise voltage is given by

$$e_n(\text{IN}) = \frac{\sqrt{(g_{m1} e_{n1})^2 + (g_{m2} e_{n2})^2}}{g_{m1} + g_{m2}}. \quad (23)$$

Generally g_{m1} and g_{m2} are taken equal so that the quiescent

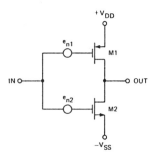

Fig. 5. CMOS push-pull amplifier.

output voltage is close to $(V_{DD} - V_{SS})/2$. In that case

$$e_n(\text{IN}) = \frac{1}{2}\sqrt{e_{n1}^2 + e_{n2}^2} = \frac{1}{2}\sqrt{\frac{a_{n1}}{Z_1 L_1} + \frac{a_{n2}}{Z_2 L_2}}. \quad (24)$$

Thus the only means to reduce noise consists in increasing the size of both transistors. On the other hand, if the transconductances of the two transistors can be taken unequal, it is possible to reduce the input noise voltage by increasing the length of the transistor having the highest a_n parameter. The equivalent input noise voltage can be written

$$e_n(\text{IN}) = \frac{\sqrt{\frac{a_{n1}}{Z_1 L_1}\left(1 + \frac{k_2' a_{n2} L_2^2}{k_1' a_{n1} L_1^2}\right)}}{1 + \sqrt{\frac{k_2' Z_2 L_1}{k_1' Z_1 L_2}}}. \quad (25)$$

V. Practical Measurements

Measurements have been made on differential and push-pull amplifiers realized with standard aluminum-gate CMOS technology. The characteristic parameters of that technology have the following values:

$$\left.\begin{array}{l} k_1' = 3~\mu A/V^2, \\ a_{n1} = 48 \times 10^3~(\mu V \times \mu m)^2 \end{array}\right\} \text{for p-channel transistors.}$$

$$\left.\begin{array}{l} k_2' = 7~\mu A/V^2, \\ a_{n2} = 380 \times 10^3~(\mu V \times \mu m)^2 \end{array}\right\} \text{for n-channel transistors.}$$

The quantities a_{n1} and a_{n2} have been obtained from rms measurements over a band from 20 Hz to 20 kHz. It appears that the n-channel transistor exhibits more noise than the p-channel transistor.

This can be due to a higher surface-state density for n-channel transistor or to the fact that the trap distribution is not uniform in the silicon bandgap, but higher near the conduction level, so that electron trapping is more efficient than hole trapping. The technological reason for this is not obvious, but its straight consequence in designing low-noise amplifiers in such a technology is that the n-channel transistor contribution to equivalent input noise voltage must be reduced.

Two differential amplifiers using p-channel transistors as input transistors, as shown in Fig. 4, have been first compared. The bias current I_D was 5 μA and the bandwidth was limited to 20 Hz - 20 kHz.

a) $Z_1 = Z_3 = 500~\mu m \quad L_1 = L_3 = 5~\mu m$

$Z_2 = Z_4 = 100~\mu m \quad L_2 = L_4 = 4~\mu m$

Measured value: $e_n(\text{IN}) = 38~\mu V$.
Calculated value: $e_n(\text{IN}) = 33.9~\mu V$.

b) $Z_1 = Z_2 = 500~\mu m \quad L_1 = L_2 = 5~\mu m$

$Z_2 = Z_4 = 50~\mu m \quad L_2 = L_4 = 44~\mu m$

Measured value: $e_n(\text{IN}) = 7.5~\mu V$.
Calculated value: $e_n(\text{IN}) = 6.9~\mu V$.

Thus, a five times noise reduction has been obtained by increasing the n-channel transistors length from 4 μm to 44 μm. It should be noted that the two amplifiers have similar voltage gains, around 44 dB.

Three push-pull amplifiers, as shown in Fig. 5, have also been compared. The current bias was 100 μA and the bandwidth limited to 20 Hz - 20 kHz.

a) $Z_1 = 1~000~\mu m \quad L_1 = 5~\mu m \quad Z_2 = 400~\mu m \quad L_2 = 4~\mu m$

b) $Z_1 = 1~000~\mu m \quad L_1 = 5~\mu m \quad Z_2 = 200~\mu m \quad L_2 = 8~\mu m$

c) $Z_1 = 500~\mu m \quad L_1 = 10~\mu m \quad Z_2 = 400~\mu m \quad L_2 = 4~\mu m$

The equivalent input noise voltages obtained from measurements are respectively 8.1 μV, 5.9 μV, and 10.5 μV and calculations from relation (25) give respectively, 7.97 μV, 5.65 μV, and 10.36 μV. These results confirm that the best noise figure is obtained with amplifier (b) where the p-channel transistor has the highest Z/L geometrical ratio and the n-channel transistor the lowest Z/L geometrical ratio. To justify that comparison, it should be noted that the transistors have the same gate area in the three cases.

Appendix

MOS device simplified equations [3] used in the saturation region, i.e., for $V_{DS} > V_{GS} - V_T$

$$I_D = k'\left(\frac{Z}{L}\right)(V_{GS} - V_T)^2$$

where

$$k' = \tfrac{1}{2}\mu C_o^*$$

and

$$V_T = V_{TO} + \gamma\left[\sqrt{V_{SB} + 2\phi_F} - \sqrt{2\phi_F}\right]$$

μ = effective carrier mobility

C_o^* = gate oxide capacitance per unit area

$$\gamma = \frac{\sqrt{2\epsilon_s q N}}{C_o^*}$$

where ϵ_s is the permittivity of Si, q the electron charge, and N the substrate impurity concentration.

Linear dependence of threshold voltage V_T versus source-substrate bias voltage V_{SB} is given by

$$\lambda \triangleq \frac{\partial V_T}{\partial V_{SB}} = \frac{\gamma}{2\sqrt{V_{SB} + 2\phi_F}}.$$

Acknowledgment

The author would like to thank C. Perrin for performing measurements.

References

[1] M. B. Das and J. M. Moore, "Measurements and interpretation of low-frequency noise is FET's," *IEEE Trans. Electron Devices*, vol. ED-21, pp. 247-257, Apr. 1974.

[2] P. Gentil, "Bruit basse fréquence du transistor MOS," *L'Onde Electrique*, vol. 58, pp. 565-575, Aug.-Sept. 1978, and pp. 645-652, Oct. 1978.

[3] Y. P. Tsividis, "Design considerations in single-channel MOS analog integrated circuits—A tutorial," *IEEE J. Solid-State Circuits*, vol. SC-13, pp. 383-391, June 1978.

… Part III: Computer-Aided Noise Analysis

Computationally Efficient Electronic-Circuit Noise Calculations

RONALD ROHRER, MEMBER, IEEE, LAURENCE NAGEL, STUDENT MEMBER, IEEE, ROBERT MEYER, MEMBER, IEEE, AND LYNN WEBER, MEMBER, IEEE

Abstract—The interreciprocal adjoint network concept is applied to the computer simulation of electronic-circuit noise performance. The method described is extremely efficient, allowing consideration of an arbitrarily large number of uncorrelated noise sources with less effort than entailed for the original small signal ac analysis. Because all noise sources may be considered, no *a priori* assumption need be made as to which noise sources are dominant in a complicated circuit. The method is illustrated with an operational-amplifier example.

INTRODUCTION

THE ELECTRICAL noise sources in passive elements and electronic devices have been investigated extensively, and appropriate models have been derived [1]–[4]. The noise performance of an electronic circuit ordinarily can be analyzed in terms of these models by considering each of the uncorrelated noise sources in turn and separately computing its contribution at the output. The overall output-noise amplitude is the square root of the sum of the squares of the amplitudes of the individual components. For simple circuits this procedure leads to expressions for total output noise, which enable the designer to assess the relative importance of the various noise sources, and thus to optimize noise performance. For a complicated circuit such as an operational amplifier, the large number of noise sources and circuit complexity completely preclude hand calculation. In fact, in ordinary usage, even machine computation of the noise contributions from all sources can be tedious, time consuming, and expensive. The usual way to avoid excessive computation in noise simulation is to attempt to isolate only the most significant noise sources (for example, those associated with input stages) and to consider only their contribution to the noise output. Aside from the obvious loss of accuracy, which arises with such an approach, there is also a reliance on the intuition of the designer, which for complicated circuits may not always be correct.

Manuscript received September 30, 1970; revised March 15, 1971. This work was sponsored by NSF Grant GK-17931 and the Joint Services Electronics Program, Grant AFOSR-68-1488.
R. Rohrer, L. Nagel, and R. Meyer are with the Department of Electrical Engineering and Computer Sciences and the Electronics Research Laboratory, University of California, Berkeley, Calif. 94720.
L. Weber was with the Department of Electrical Engineering, University of California, Berkeley, Calif. He is now with Hewlett-Packard Corp., Santa Clara, Calif.

In this paper a computational technique is described that calculates the noise contribution from an arbitrarily large number of noise sources at a given frequency with little more computer time than is normally required for a single noise source. For complicated circuits the attendant gain in computational efficiency is very large.

NOISE MODELS

The three important types of noise in electronic circuits are shot noise, thermal noise, and flicker noise. Of these noise types only the first two are to be considered here, although flicker-noise sources could be handled in the same manner were empirical data available for their characteristics. Shot noise and thermal noise are related directly to physical parameters and empirical data are not required.

The noise models for resistors, junction diodes, bipolar junction transistors, and field-effect transistors are shown in Fig. 1. Since they are to be employed here in the context of a nodal-analysis computer program, all noise sources are modeled as independent current sources. The following expressions give the noise-source amplitudes for the components shown in Fig. 1.

A. Linear Resistors

$$i_R^2 = 4kTG\Delta f \quad (1)$$

where k is Boltzman's constant, T the temperature in degrees Kelvin, G the conductance in ohms, and Δf the bandwidth under consideration.

B. Junction Diodes

$$i_{RS}^2 = 4kT(1/r_s)\,\Delta f \quad (2a)$$

$$i_D^2 = 2qI_D\,\Delta f \quad (2b)$$

where r_s is the series ohmic resistance, q the electronic charge, and I_D the direct current through the diode.

C. Bipolar Junction Transistors

$$i_{RB}^2 = 4kT(1/r_B)\,\Delta f \quad (3a)$$

$$i_B^2 = 2q(I_C/\beta_{dc})\,\Delta f \quad (3b)$$

$$i_C^2 = 2qI_C\,\Delta f \quad (3c)$$

Reprinted from *IEEE J. Solid-State Circuits*, vol. SC-6, no. 4, pp. 204–213, Aug. 1971.

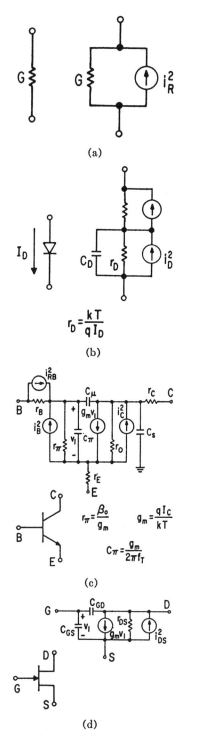

Fig. 1. Noise models for four circuit devices. (a) Linear resistors. (b) Junction diodes. (c) Bipolar junction transistors. (d) Field-effect transistors.

where r_B is the base spreading resistance in ohms, I_C the dc collector current, and β_{dc} the dc gain (in Fig. 1, β_0 is the small-signal current gain).

D. Field-Effect Transistors

$$\overline{i_{DS}^2} = 4kT(\tfrac{2}{3}g_m)\,\Delta f \tag{4}$$

where g_m is the small-signal transconductance at the operating point. The collector–base leakage current in bipolar junction transistors and the gate leakage current in field-effect transistors are both neglected (a reasonable approximation for silicon devices).

Conventional Noise Analysis

Were superposition to apply, the overall noise-simulation task would be simple, entailing merely a separate analysis with the noise sources as simultaneous driving functions. However, the noise sources are generators of random time functions with uncorrelated phases. In theory then, a separate analysis for each noise source is required to obtain the noise amplitude at the output; the desired result is the square root of the sum of the squares of the individual contributions. In other words the overall output-noise voltage is given by the rms relation

$$V_{\text{noise}} = \sqrt{\sum_{l=1}^{n} |V_l|^2} \tag{5}$$

where V_l is the output-noise voltage contributed by the lth noise source and n is the total number of noise sources in the circuit. Hence an electronic circuit with R linear resistors, D diodes, B bipolar junction transistors, and F field-effect transistors would require $R + 2D + 3D + F$ additional analyses beyond the original per frequency point for a complete noise simulation, and for even moderate-sized circuits computation time would be excessive. Throughout this discussion it is assumed that in addition to the noise analysis an original frequency response is computed, and that the results of it are available to aid in the noise computations. Even if such an approach were to be taken the situation would not be quite as bad as described above; much of the original analysis effort applies as well to the noise situation. For example, if the nodal admittance equations

$$\mathbf{YV} = \mathbf{I} \tag{6a}$$

are already solved for the node voltages

$$\mathbf{V} = \mathbf{Y}^{-1}\mathbf{I}, \tag{6b}$$

each noise source merely introduces a new forcing vector $\mathbf{I}_l$; so rather than a complete analysis only a multiplication by the already computed $\mathbf{Y}^{-1}$ need be performed

$$\mathbf{V} = \mathbf{Y}^{-1}\mathbf{I}_l \,(l = 1, 2, \cdots, n). \tag{6c}$$

Still a succession of such computations is costly in comparison to the more efficient technique to be described, which also takes similar advantage of prior analysis information.

If the circuit under simulation were reciprocal, its complete noise analysis could be performed by exciting the output with a unit source and obtaining the resultant contributions at the locations of each of the noise sources. Of course, the square root of the appropriately weighted sum of the squares of these responses must still be invoked. However, most electronic circuits of interest are nonreciprocal. The exploitation of the mutual reciprocity or interreciprocity concept allows the complete noise simulation of any electronic circuit with less additional computational effort than entailed for a single analysis.

Interreciprocity

Two linear time-invariant circuits, which are interreciprocal relative to each other exhibit the reciprocity concept between them [5]–[7]. For example, as illustrated in Fig. 2 a unit-independent current-source excitation at port **J** of one such circuit produces the same voltage response at its port **K** as a unit independent current source excitation at port **K** of the other does at its own port **J**. Given a circuit, its interreciprocal adjoint is easily constructed by means of the table of correspondences provided in Fig. 3. Both circuit topologies are identical, and so are most elements; only nonreciprocal elements are reversed. The proof that two such networks exhibit interreciprocity follows easily from Tellegen's theorem [8]:

$$\sum_{b=1}^{N_b} \hat{V}_b I_b = 0 \quad (7\mathrm{a})$$

and

$$\sum_{b=1}^{N_b} V_b \hat{I}_b = 0 \quad (7\mathrm{b})$$

where the summation is over all N_b branches in the networks, including ports, V_b and I_b are the voltage and current in branch b of network N, and $\hat{V}_b$ and $\hat{I}_b$ are the voltages and current in the corresponding branch b of network $\hat{N}$. Upon subtraction of (7b) from (7a), the resultant expression is

$$\sum_{b=1}^{N_b} (V_b \hat{I}_b - \hat{V}_b I_b) = 0. \quad (8)$$

But because of the correspondences in Fig. 3 the internal branch variables drop out and all that remains are the port variables

$$\sum_{p=1}^{N_p} (V_p \hat{I}_p - \hat{V}_p I_p) = 0 \quad (9)$$

where the summation is taken over all p ports. But for the situation shown in Fig. 2, (9) reduces to

$$V_K - \hat{V}_J = 0, \quad (10)$$

which shows the desired interreciprocity. More detailed treatments of the interreciprocity concept are given in the references [5]–[7]; the above is adequate for its use in the context of noise analysis.

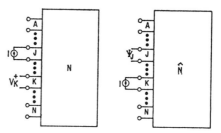

Fig. 2. If the two circuits N and $\hat{N}$ are interreciprocal, then $\Psi_J = V_K$, the remaining ports of both circuits may be either short or open circuited as long as corresponding pairs are treated identically.

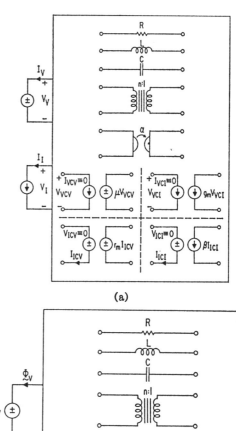

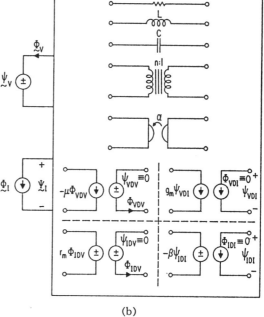

Fig. 3. Symbolic representation of (a) an arbitrary network N and (b) its interreciprocal adjoint network $\hat{N}$ by means of individual element correspondences.

Efficient Noise Computation

The complete noise simulation of any given electronic circuit could be performed in the following manner. First the correspoding interreciprocal circuit is constructed by means of the correspondence table given in Fig. 3. Fig. 4 shows for convenience the interreciprocal models for the device models originally given in Fig. 1. There are no noise sources indicated in these models since only the circuit output is to be excited. In the second step the interreciprocal circuit is analyzed at the frequency point in question with a unit current-source excitation of zero phase at its output. The resulting complex voltages produced between the nodes where the noise-independent current sources originally appeared give directly the transimpedances from the noise independent sources to theoutput in the original circuit. If these transimpedances are denoted by Z_l, then the individual contributions to the output-noise voltage magnitude are

$$V_{\text{noise}}{}^l = |Z_l I_l| \; (l = 1, 2, \cdots, n) \quad (11)$$

where I_l is the value of the lth noise-independent current source. The overall output noise is

$$V_{\text{noise}} = \sqrt{\sum_{l=1}^{n} |Z_l I_l|^2} \quad (12)$$

but now only one analysis is required regardless of the number of noise-independent current sources.

In theory then, in one analysis, plus a little additional effort, one can compute the entire noise behavior of any given electronic circuit at any given frequency point. In practice, however, the task is even easier. If network N is described by the nodal admittance matrix $\mathbf{Y}$ and network $\hat{N}$ is described by the nodal admittance matrix $\hat{\mathbf{Y}}$ and the two networks are interreciprocal, then

$$\hat{\mathbf{Y}} = \mathbf{Y}^T \quad (13)$$

where the superscript T indicates transpose. But in the original analysis at a given frequency point $\mathbf{Y}^{-1}$ must be computed, and $(\mathbf{Y}^T)^{-1} = (\mathbf{Y}^{-1})^T$, so the noise simulation becomes merely a matter of appropriate substitution with no matrix inversion involved. In any case one does not actually invert the nodal admittance matrix but merely LU factors it for even greater computational efficiency [9].

This noise-simulation scheme has been implemented in the computer analysis of nonlinear circuits, excluding radiation (CANCER) program [10], [11].

Operational-Amplifier Example

The application of the above described noise-analysis method to a complicated circuit, such as a μA 741 operational amplifier, is a good illustration of its power. The equivalent circuit schematic diagram for the μA 741 is shown in Fig. 5, where the short-circuit protection has

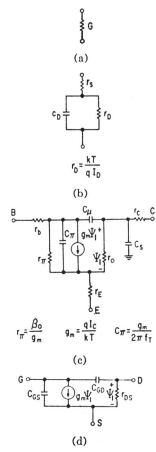

Fig. 4. Adjoint models for the four circuit devices shown in Fig. 1. (a) Linear resistors. (b) Junction diodes. (c) Bipolar junction transistors. (d) Field-effect transistors.

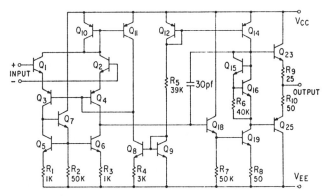

Fig. 5. Equivalent circuit schematic diagram for a μA 741 operational amplifier (short-circuit protection has been omitted from the output stage for convenience).

Fig. 6. The μA 741 operational amplifier in the circuit configuration considered in this paper.

```
*********************************************************************************************
    EECS 290   CIRCUIT 10   741 OP AMP                                        ----- CANCER -----

                    NOISE ANALYSIS                    TEMPERATURE    25.000  DEG C

*********************************************************************************************


    FREQUENCY    1.000E+00  HZ


    NOISE CONTRIBUTIONS OF RESISTORS
    NAME         EQU CURRENT        VOLTAGE
                 AMPS/RT HZ         SQ V/HZ

    RIN          1.047E-11          2.465E-16
    RF           3.312E-12          2.468E-17
    R1           4.057E-12          3.849E-15
    R2           5.737E-13          3.319E-20
    R3           4.057E-12          3.854E-15
    R4           2.342E-12          9.764E-21
    R5           8.492E-13          8.782E-23
    R6           6.414E-13          3.630E-29
    R7           5.737E-13          2.571E-20
    R8           1.814E-11          5.781E-23
    R9           2.566E-11          1.425E-28
    R10          1.814E-11          4.292E-28
    RINN         1.283E-10          1.991E-18

    NOISE CONTRIBUTIONS OF TRANSISTORS
    NAME         NOISE FROM BASE RESISTANCE     NOISE FROM BASE CURRENT        NOISE FROM COLLECTOR CURRENT
                 EQU CURRENT   VOLTAGE          EQU CURRENT   VOLTAGE          EQU CURRENT   VOLTAGE
                 AMPS/RT HZ    SQ V/HZ          AMPS/RT HZ    SQ V/HZ          AMPS/RT HZ    SQ V/HZ

    Q1           9.071E-12     3.982E-16        1.612E-13     1.125E-17        1.976E-12     2.077E-15
    Q2           9.071E-12     3.982E-16        1.613E-13     9.808E-18        1.976E-12     2.075E-15
    Q3           2.566E-11     4.978E-17        2.742E-13     4.186E-17        1.967E-12     2.201E-15
    Q4           2.566E-11     4.978E-17        2.745E-13     4.265E-17        1.969E-12     2.198E-15
    Q5           9.071E-12     7.695E-16        1.633E-13     8.670E-18        2.000E-12     3.990E-15
    Q6           9.071E-12     7.706E-16        1.633E-13     9.301E-18        2.000E-12     3.996E-15
    Q7           9.071E-12     2.528E-23        1.458E-13     4.807E-17        1.785E-12     3.214E-19
    Q8           9.071E-12     6.909E-22        2.207E-13     1.014E-22        2.878E-12     1.679E-21
    Q9           1.047E-11     4.903E-22        9.081E-13     1.812E-24        1.362E-11     7.402E-23
    Q10          2.566E-11     1.195E-21        3.368E-13     3.248E-22        2.818E-12     2.389E-20
    Q11          2.566E-11     1.195E-21        3.349E-13     3.582E-22        2.818E-12     2.528E-20
    Q12          2.566E-11     6.604E-22        2.913E-12     6.812E-24        1.303E-11     6.112E-22
    Q14          2.566E-11     6.619E-22        2.913E-12     7.137E-23        1.303E-11     7.407E-22
    Q15          9.071E-12     1.565E-27        1.841E-13     2.402E-29        2.401E-12     5.531E-27
    Q16          1.047E-11     1.256E-27        8.560E-13     8.023E-28        1.284E-11     2.670E-26
    Q18          9.071E-12     2.527E-23        1.616E-13     5.681E-17        2.107E-12     3.468E-19
    Q19          1.047E-11     1.733E-22        8.553E-13     5.782E-20        1.283E-11     4.582E-22
    Q23          0.            0.               3.028E-13     3.949E-25        5.245E-12     5.154E-27
    Q24          9.071E-12     1.149E-27        3.202E-13     1.158E-24        3.202E-12     6.304E-27


    OUTPUT NOISE VOLTAGE             1.649E-07   VOLTS/RT HZ       2.718E-14   SQ VOLTS/HZ

    EQUIVALENT INPUT NOISE VOLTAGE   1.649E-08   VOLTS/RT HZ       2.718E-16   SQ VOLTS/HZ
```

Fig. 7. (a) CANCER output at two frequency points for the μA 741 operational amplifier in the circuit of Fig. 6: 1 Hz.

been omitted from the output stage for convenience. The device model parameter values employed for the n-p-n and p-n-p bipolar junction transistors were those determined for another investigation [12] by actual probe measurements on a chip. The noise model was checked by examining the noise characteristics of individual devices from the integrated-circuit process. The measurements were in good agreement with predictions based on the independently measured values of r_B, β_{dc}, and β_0.

The particular simulation situation used to examine the noise performance of the μA 741 was that of Fig. 6. This configuration was chosen for ease of comparison between computed results and actual measurements. The voltage gain of the circuit is 10 and the effective source resistance seen by the operational amplifier is about 137 Ω. A typical CANCER output for one low-frequency point (1 Hz) is shown in Fig. 7(a). The value (in A/$\sqrt{Hz}$) of each noise source is shown in the EQU CURRENT column and the noise contribution (in V^2/Hz) that each source produces at the output is shown in the VOLTAGE column. Hence, the various noise contributions are easily compared. The unexpected result in this instance is that the resistors R_1 and R_3 and the collector-current noise generators of transistors Q_5 and Q_6 are the dominant noise sources in the circuit. These elements form the active load for the input transistors Q_1–Q_4. Note that both total output-noise voltage and equivalent input-noise voltage are also tabulated for the frequency point. The lack of a noise contribution from the

EECS 290 CIRCUIT 10 741 OP AMP ----- CANCER -----

NOISE ANALYSIS TEMPERATURE 25.000 DEG C

FREQUENCY 1.000E+06 HZ

NOISE CONTRIBUTIONS OF RESISTORS

NAME	EQU CURRENT AMPS/RT HZ	VOLTAGE SQ V/HZ
RIN	1.947E-11	2.969E-18
RF	3.312E-12	1.139E-19
R1	4.057E-12	5.437E-17
R2	5.737E-13	1.113E-18
R3	4.057E-12	5.336E-17
R4	2.347E-12	1.511E-18
R5	6.970E-13	3.228E-21
R6	6.414E-13	1.773E-19
R7	5.737E-13	2.819E-18
R8	1.814E-11	1.331E-18
R9	2.566E-11	1.137E-19
R10	1.414E-11	1.940E-19
RIN2	1.283E-13	2.711E-20

NOISE CONTRIBUTIONS OF TRANSISTORS

NAME	NOISE FROM BASE RESISTANCE		NOISE FROM BASE CURRENT		NOISE FROM COLLECTOR CURRENT	
	EQU CURRENT AMPS/RT HZ	VOLTAGE SQ V/HZ	EQU CURRENT AMPS/RT HZ	VOLTAGE SQ V/HZ	EQU CURRENT AMPS/RT HZ	VOLTAGE SQ V/HZ
Q1	9.071E-12	4.423E-18	1.512E-13	9.150E-20	1.974E-12	2.190E-17
Q2	9.071E-12	4.097E-18	1.613E-13	1.257E-19	1.976E-12	2.087E-17
Q3	2.566E-11	5.603E-19	2.782E-13	4.094E-19	1.967E-12	6.834E-17
Q4	2.556E-11	5.342E-19	2.785E-13	5.692E-19	1.969E-12	4.600E-17
Q5	9.071E-12	1.065E-17	1.633E-13	1.765E-19	2.000E-12	6.838E-17
Q6	9.071E-12	1.106E-17	1.633E-13	2.882E-19	2.000E-12	4.701E-17
Q7	9.071E-12	1.946E-19	1.459E-13	7.037E-19	1.785E-12	1.074E-17
Q8	9.071E-12	1.013E-19	2.207E-13	1.571E-20	2.878E-12	3.117E-19
Q9	1.047E-11	7.774E-20	9.091E-13	2.870E-22	1.362E-11	1.189E-20
Q10	2.566E-11	7.262E-20	3.358E-13	9.901E-21	2.818E-12	7.297E-19
Q11	2.566E-11	7.591E-20	3.369E-13	1.094E-20	2.818E-12	3.963E-18
Q12	2.556E-11	2.601E-19	2.913E-12	1.334E-20	1.303E-11	1.535E-18
Q14	2.556E-11	7.535E-19	2.913E-12	2.074E-19	1.303E-11	8.710E-18
Q15	9.071E-12	7.783E-19	1.841E-13	1.172E-20	2.401E-12	2.702E-18
Q16	1.047E-11	6.239E-19	8.560E-13	3.916E-19	1.284E-11	2.657E-19
Q18	9.071E-12	4.692E-19	1.616E-13	9.139E-19	2.107E-12	3.799E-17
Q19	1.047E-11	3.983E-18	8.553E-13	7.388E-19	1.283E-11	3.203E-18
Q23	0.	0.	3.028E-13	8.743E-22	5.245E-12	5.716E-19
Q24	9.071E-12	9.097E-19	5.217E-13	3.396E-21	5.202E-12	6.790E-19

OUTPUT NOISE VOLTAGE 2.293E-08 VOLTS/RT HZ 5.213E-16 SQ VOLTS/HZ

EQUIVALENT INPUT NOISE VOLTAGE 2.083E-08 VOLTS/RT HZ 4.339E-16 SQ VOLTS/HZ

Fig. 7. (b) CANCER output at two frequency points for the μA 741 operational amplifier in the circuit of Fig. 6: 1 MHz.

base resistance of Q_{23} is due to the assumption of a zero value for its base resistance because of lack of data for this device; this device is an output transistor, and its contribution to the total noise is negligible.

For comparison with the above, an output at a higher frequency (1 MHz) is shown in Fig. 7(b). At this frequency the equivalent input-noise voltage has risen from 16.5 to 20 nV/$\sqrt{Hz}$. It is interesting to note the changes in the relative significance of the various noise generators. Resistors R_1 and R_1 and the collector-current noise generators of Q_5 and Q_6 are still among the most significant noise sources. However, the collector-current generators of Q_3 and Q_4 have become more important and Q_1 and Q_2 relatively less important. Transistor Q_{18}, which was previously completely negligible as a noise source, is now quite significant due to the attenuation of the 30-pF frequency-shaping capacitor.

The program output also gives a separate listing of the total noise referred to the input at the specified frequencies, together with the circuit gain. These data are available in graphical form as shown in Fig. 8(a)–(c), where graphs are presented of total output noise, equivalent input noise and circuit gain as a function of frequency for this μA 741 example. For the μA 741 the equivalent input noise is essentially constant up to 0.5 MHz (with flicker noise neglected) at a value of 16.5 nV/$\sqrt{Hz}$. It is interesting to note that this is the noise in a 16 kΩ resistor giving the μA 741 a noise figure of approximately 20 dB in the circuit shown.

The program can be used to compute the noise per-

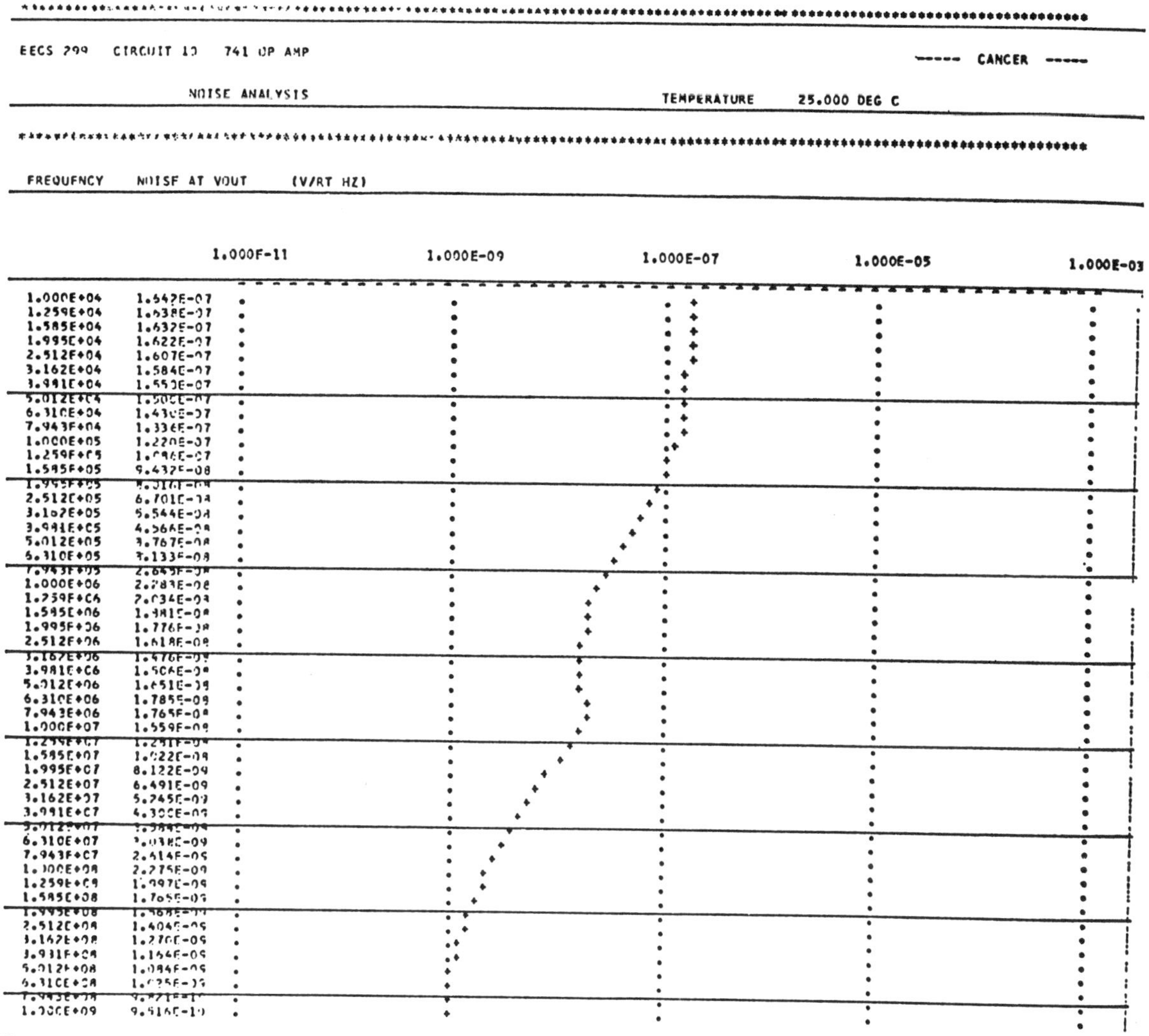

Fig. 8. (a) Graphical output from CANCER for the μA 741 operational amplifier in the circuit of Fig. 6: output noise voltage spectral density.

```
EECS 299   CIRCUIT 10   741 OP AMP                                          ----- CANCER -----
              NOISE ANALYSIS                        TEMPERATURE    25.000 DEG C

 FREQUENCY    INPUT NOISE REFERRED FROM VOUT    (V/RT HZ)

                     1.585E-08         2.512E-08         3.981E-08         6.310E-08         1.000E-07
 1.000E+04   1.649E-08
 1.259E+04   1.649E-08
 1.585E+04   1.649E-08
 1.995E+04   1.649E-08
 2.512E+04   1.649E-08
 3.162E+04   1.649E-08
 3.981E+04   1.649E-08
 5.012E+04   1.650E-08
 6.310E+04   1.650E-08
 7.943E+04   1.652E-08
 1.000E+05   1.653E-08
 1.259E+05   1.656E-08
 1.585E+05   1.661E-08
 1.995E+05   1.668E-08
 2.512E+05   1.679E-08
 3.162E+05   1.695E-08
 3.981E+05   1.724E-08
 5.012E+05   1.767E-08
 6.310E+05   1.835E-08
 7.943E+05   1.936E-08
 1.000E+06   2.083E-08
 1.259E+06   2.300E-08
 1.585E+06   2.630E-08
 1.995E+06   3.103E-08
 2.512E+06   4.201E-08
 3.162E+06   5.361E-08
 3.981E+06   5.657E-08
 5.012E+06   5.674E-08
 6.310E+06   5.725E-08
 7.943E+06   5.727E-08
 1.000E+07   5.583E-08
 1.259E+07   5.263E-08
 1.585E+07   4.816E-08
 1.995E+07   4.327E-08
 2.512E+07   3.870E-08
 3.162E+07   3.484E-08
 3.981E+07   3.182E-08
 5.012E+07   2.954E-08
 6.310E+07   2.786E-08
 7.943E+07   2.665E-08
 1.000E+08   2.585E-08
 1.259E+08   2.545E-08
 1.585E+08   2.549E-08
 1.995E+08   2.592E-08
 2.512E+08   2.671E-08
 3.162E+08   2.775E-08
 3.981E+08   2.893E-08
 5.012E+08   3.013E-08
 6.310E+08   3.124E-08
 7.943E+08   3.223E-08
 1.000E+09   3.311E-08
```

Fig. 8. (b) Graphical output from CANCER for the µA 741 operational amplifier in the circuit of Fig. 6: equivalent input-noise voltage spectral density.

formance for any value of source impedance. Either equivalent input noise voltage or noise current can be emphasized by using either a low or a high value, respectively, of source resistance. If noise representation by an equivalent input current noise generator is of interest for a particular source resistance, it can be obtained directly by specifying a current source input for the small-signal ac solution.

The computer results described above for the circuit of Fig. 6 were checked experimentally. The output-noise voltage in a 200-Hz bandwidth and the voltage gain were measured at a number of frequencies, and the equivalent input-noise voltage per unit bandwidth was thus determined. These measurements were performed on a number of samples of the µA 741 and the results are shown in Fig. 9 together with the computer predictions. At higher frequencies the agreement between the simulated and measured noise outputs is excellent; the deviations at low frequencies is the result of the omission of flicker-noise sources from the program. The noise levels shown in Fig. 9 are in close agreement with manufacturer's specifications for the equivalent input-noise voltage of the µA 741. The spread in experimental values is attributable to variations in the bias currents in transistors Q_1–Q_6.

Table I shows the execution times for both conventional noise simulation and adjoint noise analysis for several circuits. The µA 741 operational amplifier discussed in this paper is the ninth circuit listed in Table I. All CANCER execution times are for the University of California, Berkeley, CDC 6400 computer and include input–output time. For the largest circuit tested, the adjoint method required approximately 30 percent additional execution time, while the conventional noise analysis required an order of magnitude additional execution time.

CONCLUSIONS

The adjoint network concept has made electronic-circuit noise simulation a virtual bonus in the overall computational analysis. Because every noise-source con-

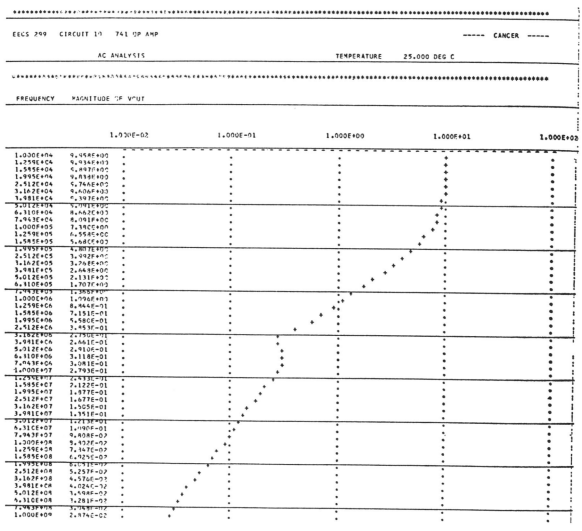

Fig. 8. (c) Graphical output from CANCER for the μA 741 operational amplifier in the circuit of Fig. 6: voltage gain versus frequency.

TABLE I
NOISE SIMULATION COMPARISON TIMES[a]

Circuit	Nodes	Q	ac[b]	ac[c] Adjoint	ac[d] Conventional
1	7	2	3.7	4.0	6.2
2	14	4	4.6	5.3	12.3
3	19	5	5.4	6.3	18.2
4	20	16	6.7	8.3	95.0
5	28	8	6.4	7.8	26.9
6	32	8	7.7	9.5	42.0
7	38	11	7.5	9.4	57.3
8	40	10	8.3	10.5	—
9	54	15	19.6	18.0	151.0
10	55	19	13.3	16.6	165.0

[a] All run times are on the CDC 6400 computer, University of California, Berkeley, and include input–output time.
[b] Time required to determine dc operating point and ac small-signal response at 100 frequency points.
[c] Time required to determine dc operating point, ac small signal response, and noise response at 100 frequency points using adjoint approach.
[d] Time required to determine dc operating points, ac small-signal response, and noise response at 100 frequency points using conventional noise analysis.

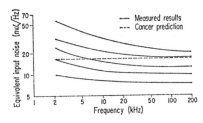

Fig. 9. Graphical comparison between measured and CANCER-predicted equivalent input-noise voltage for the µA 741 operational amplifier in the circuit configuration of Fig. 6.

tribution may be efficiently accounted for, there is obvious application to the noise performance optimization of linear electronic circuits.

References

[1] D. G. Peterson, "Noise performance of transistors," *IRE Trans. Electron Devices*, vol. ED-4, May 1962, pp. 296–303.
[2] A. Van der Ziel, *Fluctuation Phenomena in Semiconductors*. London: Butterworth, 1959.
[3] ——, "Thermal noise in field-effect transistors," *Proc. IRE*, vol. 50, Aug. 1962, pp. 1808–1812.
[4] A. G. Jordan and N. A. Jordan, "Theory of noise in metal-oxide semiconductor devices," *IEEE Trans. Electron Devices*, vol. ED-12, Mar. 1965, pp. 148–156.
[5] J. L. Bordewijk, "Inter-reciprocity applied to electrical networks," *Appl. Sci. Res.*, sect. B, vol. 6, 1956, pp. 61–74.
[6] S. W. Director and R. A. Rohrer, "Inter-reciprocity and its Implications," *Proc. Int. Symp. Network Theory* (Belgrad), Sept. 1968, pp. 11–30.
[7] P. Penfield, Jr., R. Spence, and S. Duinker, "A generalized form of Tellegen's theorem," *IEEE Trans. Circuit Theory*, vol. CT-17, Aug. 1970, pp. 302–305.
[8] C. A. Desoer and E. S. Kuh, *Basic Circuit Theory*. New York: McGraw-Hill, 1969, ch. 9, sect. 4, pp. 392–396.
[9] S. W. Director, "*LU* factorization and network sensitivities," *IEEE Trans. Circuit Theory* (Corresp.), vol. CT-18, Jan. 1971, pp. 3–10.
[10] R. A. Rohrer, L. W. Nagel, R. Meyer, and L. Weber, "CANCER: Computer analysis of nonlinear circuits, excluding radiation," presented at *1971 ISSCC*, Philadelphia, Pa., Feb. 18, 1971.
[11] L. Nagel and R. Rohrer, "Computer analysis of nonlinear circuits, excluding radiation—CANCER," *IEEE J. Solid-State Circuits*, this issue pp. 166–182.
[12] B. A. Wooley and D. O. Pederson, "A computer-aided evaluation of new integrated circuit operational amplifiers," *IEEE NEREM Rec.*, 1969, pp. 88–89.

Network Sensitivity and Noise Analysis Simplified

FRANKLIN H. BRANIN, JR.

Abstract—Formulas are easily derived for computing the sensitivity of all the response variables of a network with respect to variation of a single parameter, and the computations can be carried out very efficiently. The converse problem of computing the sensitivity of a single response variable with respect to variations of several parameters, though apparently more difficult, has recently been solved with equal efficiency by appealing to the concept of an "adjoint network." This concept, however, is shown here to be superfluous, since equivalent (and slightly simpler) formulas can be derived using well-known matrix manipulations alone.

The problem of computing the signal/noise ratio of a single network response variable has also been solved by using the adjoint network concept. But, here again, standard matrix manipulations suffice to yield the same results with less conceptual encumbrance. Thus the adjoint network approach, though still valid, proves to be unnecessary for solving these two problems.

Introduction

A GREAT DEAL of interest has been aroused by the application of the adjoint network concept to the problems of network sensitivity [1]–[3] and noise analysis [4]. This concept [5], [6], which is based on an ingenious use of Tellegen's theorem [7], [8], is employed in the derivation of very efficient algorithms that permit the sensitivities or noise responses of a network to be computed with only slightly more effort than is required for the original analysis of the network.

In the author's opinion, however, the adjoint network is an unnecessarily complicated basis for deriving these algorithms. Indeed, fully equivalent and somewhat simpler algorithms can be derived without appealing to the adjoint network concept at all. The resulting computations involve little more than a virtual matrix transposition and a few matrix-vector multiplications. Moreover, the new algorithms are free of a self-cancelling pair of sign changes which are inherent in the adjoint network approach.

This simplified derivation in no way impugns the validity of the adjoint network concept. But it does help to put the theories of network sensitivity and noise analysis into better perspective by providing an alternative way of deriving the pertinent equations.

Basis of the Derivation

The basic matrix operations involved in deriving both the sensitivity and noise analysis algorithms can be illustrated as follows. We consider the system of linear equations

$$AX = B \qquad (1)$$

Manuscript received May 9, 1972; revised September 11, 1972.
The author is with the IBM Corporation, System Product Division, Kingston, N. Y.

where A is an $n \times n$ nonsingular matrix and both X and B are $n \times m$ matrices instead of vectors. The formal solution of (1) is, of course,

$$X = A^{-1}B. \qquad (2)$$

Now, if we are interested in obtaining only the jth column of the X-matrix, we can postmultiply both sides of (2) by the elementary column vector e_j which contains $+1$ in the jth row and zeroes everywhere else. Thus we obtain

$$X_{\cdot j} = X e_j = A^{-1} B e_j = A^{-1} B_{\cdot j} \qquad (3)$$

where $X_{\cdot j}$ and $B_{\cdot j}$ are the jth columns of X and B. Similarly, if we are interested in only the ith row of X, we can premultiply by e_i^t, obtaining

$$X_{i \cdot} = e_i^t X = e_i^t A^{-1} B = y_i^t B. \qquad (4)$$

Here, y_i^t is the ith row of the inverse matrix A^{-1}, or

$$y_i^t = e_i^t A^{-1}. \qquad (5)$$

But (5) is equivalent to

$$A^t y_i = e_i \qquad (6)$$

which shows that y_i can be computed using the transpose of the original coefficient matrix A. $X_{i \cdot}$ can then be computed as the product of y_i^t and B. This simple operation is the essence of the derivations to follow.

Sensitivity Calculations

We are concerned with calculating what we shall call "column" sensitivity and "row" sensitivity vectors. The first of these involves the sensitivities of many network responses with respect to a *single* parameter while the second involves the sensitivity of a single network response with respect to *many* parameters.

Both types of sensitivity are important and they can be derived using matrix manipulations similar to those described for obtaining the columns and rows of the X-matrix in the previous section. To illustrate the derivation, we shall base our discussion on the hybrid formulation of the network equations as used in the ECAP II program [9]–[11].

The Kirchhoff laws and Ohm's law for a dc network can be expressed by the equations

$$w = Tx + Sf \qquad (7)$$
$$v = H(w)w + v'(w) \qquad (8)$$
$$0 = T^t v + Mx + Wf \qquad (9)$$

where T, S, M, and W are topological matrices, $H(w)$ is the hybrid-parameter matrix (possibly nonlinear), and

Reprinted from *IEEE Trans. Circuit Theory*, vol. CT-20, no. 3, pp. 285–288, May 1973.

the vectors w, x, f, v, and $v'(w)$ have the following meanings:

$$w = \begin{bmatrix} V_G \\ I_R \end{bmatrix} \quad (10)$$

where V_G and I_R are the voltages across conductance branches and current through resistance branches;

$$x = \begin{bmatrix} V_{TG} \\ I_{LR} \end{bmatrix} \quad (11)$$

where V_{TG} and I_{LR} refer to conductance tree-branch voltages and resistance link currents;

$$f = \begin{bmatrix} E \\ J \end{bmatrix} \quad (12)$$

where E and J are the independent voltage and current sources;

$$v = \begin{bmatrix} I_G \\ V_R \end{bmatrix} \quad (13)$$

where I_G and V_R are the dual of the w-vector; and

$$v'(w) = \begin{bmatrix} E(w) \\ J(w) \end{bmatrix} \quad (14)$$

where $E(w)$ and $J(w)$ are the dependent voltage and current sources—possibly nonlinear.

Now if the network elements are a function of a single parameter p, then differentiation of (7)–(9) yields

$$\frac{dw}{dp} = T\frac{dx}{dp} + S\frac{df}{dp} \quad (15)$$

$$\frac{dv}{dp} = \left(\frac{\partial v}{\partial w}\right)_p \frac{dw}{dp} + \left(\frac{\partial v}{\partial p}\right)_w \quad (16)$$

$$0 = T^t\frac{dv}{dp} + M\frac{dx}{dp} + W\frac{df}{dp} \quad (17)$$

which may be combined to give the expression

$$\left[T^t\left(\frac{\partial v}{\partial w}\right)_p T + M\right]\frac{dx}{dp}$$
$$= -\left[T^t\left(\frac{\partial v}{\partial w}\right)_p S + W\right]\frac{df}{dp} - T^t\left(\frac{\partial v}{\partial p}\right)_w. \quad (18)$$

This equation may be solved directly for the dx/dp-vector (or column sensitivity vector) by the LU decomposition of the Jacobian matrix $[T^t(\partial v/\partial w)_p T + M]$ since the right-hand-side vector can be evaluated explicitly. Normally, the LU factors of the Jacobian matrix will already have been computed in the course of performing the dc analysis of the network. Therefore, once the right-hand side of (18) has been evaluated, little more than a backsubstitution process (using the LU factors) is required to compute the column sensitivity vector dx/dp—and with the aid of (15) and (16), dw/dp and dv/dp also.

Suppose, however, that the network elements are a function of many parameters, all of which are included in a p-vector. Then (15)–(18) still apply, but with the quantities dw/dp, dx/dp, df/dp, and dv/dp now being matrices instead of vectors. We can, of course, solve (15) for the entire dx/dp-matrix, one column at a time. But this is too costly, particularly if we happen to be interested in only one row of dx/dp, that is, a *row* sensitivity vector. By recourse to the manipulations outlined in (1)–(6), we can obtain a row sensitivity vector about as easily as a column sensitivity vector.

For example, if we desire the sensitivity of the response variable x_i alone with respect to all the elements of the p-vector, that is the ith row of the dx/dp-matrix, we must first solve the equation

$$\left[T^t\left(\frac{\partial v}{\partial w}\right)_p T + M\right]^t y_i = e_i \quad (19)$$

analogous to (6). Then, patterning (4), we premultiply the entire right-hand-side matrix of (18) by y_i^t to find the desired row sensitivity vector.

The ith row sensitivity vector in the matrix dw/dp can be found almost as simply. We premultipy (15) by e_i^t to yield

$$e_i^t \frac{dw}{dp} = (e_i^t T)\frac{dx}{dp} + e_i^t S \frac{df}{dp}. \quad (20)$$

Then, letting $y_i^t = e_i^t T$ represent the ith row of the T-matrix, we solve the equation

$$\left[T^t\left(\frac{\partial v}{\partial w}\right)_p T + M\right]^t z_i = y_i = T^t e_i \quad (21)$$

in place of (19). The next step is to premultiply the right-hand side of (18) by z_i^t and finally add to this the second term on the right-hand side of (20).

To calculate the ith row sensitivity vector in the matrix dv/dp, we write

$$e_i^t \frac{dv}{dp} = \left[e_i^t\left(\frac{\partial v}{\partial w}\right)\right]\frac{dw}{dp} + e_i^t\left(\frac{\partial v}{\partial p}\right)_w. \quad (22)$$

This time, letting $y_i^t = e_i^t(\partial v/\partial w)_p$ represent the ith row of the matrix $(\partial v/\partial w)_p$, we premultiply (15) by y_i^t to obtain the result

$$y_i^t \frac{dw}{dp} = (y_i^t T)\frac{dx}{dp} + y_i^t S \frac{df}{dp}. \quad (23)$$

Accordingly, in place of (21), we must now solve

$$\left[T^t\left(\frac{\partial v}{\partial w}\right)_p T + M\right]^t z_i = T^t y_i. \quad (24)$$

Finally, we premultiply the right-hand side of (18) by z_i^t and add to the result the second terms on the right-hand sides of both (23) and (22).

Common to the task of computing any of these row sensitivity vectors is the necessity to compute the auxiliary sensitivity matrices df/dp and $(\partial v/\partial p)_w$. These matrices may not be trivial to calculate but they are likely to be sparse enough for their computational cost to be small compared to an entire dc analysis. The matrix $(\partial v/\partial w)_p$, incidentally, is already available since it is required by the dc analysis.

Assuming, as is true in the case of ECAP II, that the LU factors of the Jacobian matrix are also available from the dc analysis, the transposes of the L and U factors can be used in solving (19), (21), and (24). Indeed, actual transposition of the L and U factors themselves can be avoided by modifying the backsubstitution process, which uses these factors, so that the effect of using their transposes is achieved. This is a straightforward programming task. Thus "virtual" transposition of the Jacobian matrix can be achieved at no additional cost in machine time. On the other hand, since row premultiplications of df/dp, $(\partial v/\partial w)_p$, and $(\partial v/\partial p)_w$ are required, these matrices would either have to be transposed or premultiplication routines would have to be provided. In any event, the expense of computing either a row or a column sensitivity vector is much less than that of a single dc analysis.

If the adjoint network approach is used for computing row sensitivity vectors (it is, of course, superfluous for column sensitivity vectors), the virtual transposition of the L and U factors cannot be used intact. A sign change affecting certain off-diagonal elements of the L and U matrices must be taken into account in addition to the transposition. This can most easily be seen in terms of a linear network problem, in which case (8) takes the simple form $v = Hw$, or using (10) and (13),

$$\begin{bmatrix} I_G \\ V_R \end{bmatrix} = \begin{bmatrix} G & F \\ N & R \end{bmatrix} \begin{bmatrix} V_G \\ I_R \end{bmatrix}. \quad (25)$$

Here, the submatrix F refers to "transfluences" [10], or current-controlled current sources, while N refers to "transpotentials" [10], or voltage-controlled voltage sources.

The corresponding adjoint network, for which the topological matrices T, M, S, and W are identical, has as its hybrid parameter matrix

$$\overline{H} = \begin{bmatrix} G^t & -N^t \\ -F^t & R^t \end{bmatrix} \quad (26)$$

which differs from the transpose of H by having the signs of its off-diagonal submatrices reversed [12]. Since these sign reversals propagate, the corresponding off-diagonal submatrices in the Jacobian matrix of the adjoint network as well as in the related L and U factors are also reversed in sign [12]. Therefore, these sign changes must be incorporated in the modified backsubstitution process along with the virtual transposition discussed above.

These sign reversals are not only a programming (and conceptual) nuisance; they are entirely vacuous since they are compensated by sign reversals in the corresponding subvectors in the adjoint counterparts of (19), (21), and (24). Consequently, the present algorithms, which are both conceptually simpler and quite free of this inconsequential sign change, give results identical to those derived by means of the adjoint network.

AC NOISE ANALYSIS

The foregoing derivation may be readily extended to the problem of noise analysis in an ac network. Equations (7) and (9), with complex vectors w, x, f, and v but with real (topological) matrices T, M, S, and W apply directly. However, (8), even in the case of small signal ac analysis of a nonlinear network at the dc operating point [13], assumes the simpler (linear) form of (25). Combining these equations, we obtain the relation

$$[T^t HT + M]x = -[T^t HS + W]f \quad (27)$$

where the bracketed matrices as well as the x- and f-vectors are complex.

The problem of noise analysis consists of computing the signal/noise ratio of one or more network response variables when several noise sources are present in addition to the signal source(s). Assuming that there are k different noise sources, let $f^{(i)}$, for $i = 1, 2, 3, \cdots, k$, represent the f-vectors corresponding to each noise source in turn being nonzero, with all other sources zero. Also, let $f^{(k+1)}$ represent the f-vector corresponding to the signal source(s) being nonzero, with all the noise sources zero.

The apparent task, then, is to solve (27) with $k+1$ different right-hand sides, or with x and f being regarded as matrices having $k+1$ columns each. But this brute force solution is too costly if we are interested in the signal/noise ratio of only one (or possibly a few) of the network response variables. Accordingly, we may resort once more to the technique outlined in (1)–(6).

Specifically, to compute the signal/noise ratio of the ith element of the x-vector, we solve

$$[T^t HT + M]^t y_i = e_i \quad (28)$$

analogous to (19) and then premultiply the right-hand side of (27) by $y_i{}^t$. Denoting the resulting row vector (of dimension $k+1$) as r, we may define the desired signal/noise ratio by the expression

$$S/N = \sqrt{r_{k+1}{}^2 \Big/ \sum_{i=1}^{k} r_i{}^2}. \quad (29)$$

To compute the signal/noise ratio of the ith element of the w-vector, we note that the relation

$$e_i{}^t w = (e_i{}^t T)x + e_i Sf \quad (30)$$

implies that we must solve the equation

$$[T^t HT + M]^t z_i = T^t e_i \quad (31)$$

analogous to (21). We then premultiply the right-hand side of (27) by z_i^t and finally add the second term of the right-hand side of (30). Equation (29) may then be applied.

If the signal/noise ratio of the ith element of the v-vector is required, then the relations

$$e_i^t v = (e_i^t H)w = y_i^t w \tag{32}$$

and

$$y_i^t w = (y_i^t T)x + y_i^t Sf \tag{33}$$

indicate that in place of (31) we must solve

$$[T^t H T + M]^t z_i = T^t y_i = T^t H^t e_i. \tag{34}$$

Next, we premultiply the right-hand side of (27) by z_i^t and last, we add the second term on the right-hand side of (33), before using (29).

Clearly, these algorithms permit the calculation of signal/noise ratios with significantly less additional work than is required for a single ac analysis.

Relation to Other Formulations

We have illustrated the matrix derivation of sensitivity and noise analysis with reference only to the hybrid formulation, which, of course, includes both the mesh and cutset formulations as special cases. But the derivation is equally applicable to any other formulation, such as the nodal or sparse tableau method.

To show this, we may write the network equations in the general form

$$g(x, f, p) = 0 \tag{35}$$

where g is a vector function of the (unknown) response vector x, the driving force vector f, and the parameter vector p. It follows that

$$\frac{dg}{dp} = \left(\frac{\partial g}{\partial x}\right)\frac{dx}{dp} + \left(\frac{\partial g}{\partial f}\right)\frac{df}{dp} + \left(\frac{\partial g}{\partial p}\right) = 0 \tag{36}$$

or

$$\left(\frac{\partial g}{\partial x}\right)\frac{dx}{dp} = -\left(\frac{\partial g}{\partial f}\right)\frac{df}{dp} - \left(\frac{\partial g}{\partial p}\right) \tag{37}$$

where dx/dp is the sensitivity matrix. Since (37) has the same form as (1), it is clear that column and/or row sensitivities can always be calculated using the manipulations prescribed by (3)–(6). Only the details of these manipulations will change from one method of formulation to another. A similar argument applies, of course, to noise analysis.

Conclusions

1) Using the standard matrix operations of transposition and multiplication by elementary vectors, computationally efficient algorithms can be derived for obtaining network sensitivities and signal/noise ratios using any method of formulating the network equations.

2) These algorithms yield results which are identical to those derived through the use of the adjoint network concept. Moreover, the present derivation has the conceptual advantage of simplicity and a slight programming advantage of freedom from the self-compensating sign reversals that are associated with certain elements of the adjoint network.

3) The adjoint network concept is not invalidated by the present work, but it is shown to be unnecessary.

References

[1] S. W. Director and R. A. Rohrer, "Inter-reciprocity and its implications," in *Proc. Int. Symp. Network Theory*, pp. 11–30, 1968.
[2] ——, "The generalized adjoint network and network sensitivities," *IEEE Trans. Circuit Theory*, vol. CT-16, pp. 318–323, Aug. 1969.
[3] ——, "Automated network design—The frequency-domain case," *IEEE Trans. Circuit Theory*, vol. CT-16, pp. 330–337, Aug. 1969.
[4] R. Rohrer, L. Nagel, R. Meyer, and L. Weber, "Computationally efficient electronic-circuit noise calculations," *IEEE J. Solid-State Circuits*, vol. SC-6, pp. 204–213, Aug. 1971.
[5] G. D. Hachtel, "Two classes of variational networks and their exploitation in computer aided design," in *Proc. 5th Annu. Allerton Conf. Circuit and Systems Theory*, pp. 564–571, 1967.
[6] R. A. Rohrer, "Synthesis of arbitrarily tapered lossy transmission lines," in *Proc. Symp. Generalized Networks*, vol. 16, pp. 115–136, 1966.
[7] B. D. H. Tellegen, "A general network theorem, with applications," *Phillips Res. Rep.*, vol. 7, pp. 259–269, 1952.
[8] P. Penfield, R. Spence, and S. Duinker, *Tellegen's Theorem and Electrical Networks*. Cambridge, Mass.: M.I.T. Press, 1970.
[9] F. H. Branin and L. E. Kugel, "The hybrid method of network analysis," presented at the IEEE Int. Symp. Circuit Theory, San Francisco, Calif., 1969.
[10] F. H. Branin, Jr., G. R. Hogsett, R. L. Lunde, and L. E. Kugel, "ECAP II—A new electronic circuit analysis program," *IEEE J. Solid-State Circuits*, vol. SC-6, pp. 146–166, Aug. 1971.
[11] *ECAP-II Program Description Manual*, IBM Corp., White Plains, N. Y., SH20-1015-0.
[12] R. N. Gadenz and G. C. Temes, "Efficient hybrid and state-space analysis of the adjoint network," *IEEE Trans. Circuit Theory* (Corresp.), vol. CT-19, pp. 520–521, Sept. 1972.
[13] F. H. Branin, Jr., and G. Martin, "Small signal ac analysis of nonlinear circuits," in *Proc. IEEE Conf. Systems, Networks, and Computers* (Oaxtepec, Mex.), pp. 516–518, 1971.

An Efficient Method for Computer Aided Noise Analysis of Linear Amplifier Networks

HERBERT HILLBRAND AND PETER H. RUSSER

Abstract—A method for computer aided noise analysis is presented which is based on a description of noise by means of correlation matrices. The method is a two-port analysis and it is, therefore, applicable to circuits which are composed of simple two-ports with known noise performance. The correlation matrix concept holds two main advantages over other methods of noise analysis. Partially correlated noise sources can be treated without any loss of efficiency and information concerning minimum noise figure and noise matching conditions is obtained.

I. INTRODUCTION

IN RECENT YEARS, several methods for computer aided noise analysis of linear networks have been reported [1]–[4]. The basic concept which is common to all these methods consists in replacing noise sources by nonrandom sinusoidal sources of the same available power and then applying a straightforward ac analysis for the noise power calculations. The advantages of this concept are obvious: noise power and noise figure calculations can be carried out with any existing ac analysis program. However, the efficiency of the method is restricted to applications where noise sources occurring in the circuit are either uncorrelated or fully correlated. If partially correlated sources are involved serious difficulties will arise, since the whole information about correlation gets lost when noise sources are replaced by nonrandom sources. In principle, the difficulties can be overcome by either two methods, first by transforming the noise network to its canonical form where only uncorrelated sources are contained and second by establishing a set of noise sources which are either uncorrelated or fully correlated and then representing each noise source occurring in the network by a linear combination of this set. Both methods are tedious and time consuming and make the analysis concept unefficient for circuits containing partially correlated sources (e.g., circuits containing any type of transistors). As another weakness of this analysis concept, no information is provided about minimum noise figure and noise matching conditions. In the opinion of the authors, this is also an important disadvantage. There are many applications where this information is absolutely required.

It is the purpose of this paper to present quite another approach to computer aided noise analysis which does not show the disadvantages of the above mentioned methods. It is based on the circuit theory of linear noisy networks. In this theory, noise in linear circuits is described in terms of correlation matrices rather than by voltages and currents. The properties of such a description are usually presented in rather general form and so the theory is not directly applicable to computer aided noise analysis. This may be one reason why the correlation matrix approach has not been used so far.

The noise analysis will be a two-port analysis. The philosophy behind such a method is as follows. The two-port which is to be analyzed is viewed as an interconnection of basic two-ports whose noise behavior is known. Starting from these basic two-ports, the analysis proceeds by interconnecting simpler two-ports to more complicated two-ports until finally the noise performance of the original two-port is obtained. During the whole analysis only two-ports are involved. Therefore, the following theoretical considerations concerning the correlation matrices are applied to two-ports only.

Obviously, the two-port analysis concept is applicable only to a certain class of networks. However, this restriction is not too severe since most practical networks, particularly in the high-frequency and microwave region, belong to this class. On the other hand, the procedure is very simple, easy to program and represents a basis for investigations towards a more general approach.

II. CORRELATION MATRIX REPRESENTATION OF NOISY TWO-PORTS

The circuit theory of linear noisy networks shows that any noisy two-port can be replaced by a noise equivalent circuit which consists of the original two-port (now assumed to be noiseless) and two additional noise sources [5]. In general, six different forms of noise equivalent circuits exist depending upon the type of the additional noise sources and their arrangement relative to the noiseless two-port. Each form is called a representation. For common applications only three of these representations

---

Manuscript received December 26, 1974; revised November 10, 1975. This work was partially supported by the Ministry of Research and Technology, Germany. The authors alone are responsible for the contents.

The authors are with Aeg-Telefunken, Forschungsinstitut, D-79 Ulm/Donau, Germany.

Reprinted from *IEEE Trans. Circuits Syst.*, vol. CAS-23, no. 4, pp. 235–238, Apr. 1976.

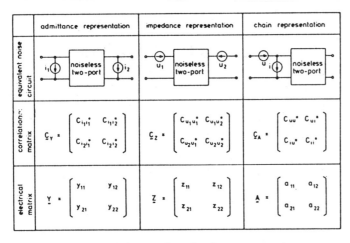

Fig. 1. Correlation matrices of various representations.

TABLE I
TRANSFORMATION MATRICES

		original representation		
		admittance	impedance	chain
resulting representation	admittance	$\begin{bmatrix} 1 & 0 \\ 0 & 1 \end{bmatrix}$	$\begin{bmatrix} y_{11} & y_{12} \\ y_{21} & y_{22} \end{bmatrix}$	$\begin{bmatrix} -y_{11} & 1 \\ -y_{21} & 0 \end{bmatrix}$
	impedance	$\begin{bmatrix} z_{11} & z_{12} \\ z_{21} & z_{22} \end{bmatrix}$	$\begin{bmatrix} 1 & 0 \\ 0 & 1 \end{bmatrix}$	$\begin{bmatrix} 1 & -z_{11} \\ 0 & -z_{21} \end{bmatrix}$
	chain	$\begin{bmatrix} 0 & a_{12} \\ 1 & a_{22} \end{bmatrix}$	$\begin{bmatrix} 1 & -a_{11} \\ 0 & -a_{21} \end{bmatrix}$	$\begin{bmatrix} 1 & 0 \\ 0 & 1 \end{bmatrix}$

are required. They are shown in Fig. 1. The additional noise sources are indicated by circles. The admittance representation uses two current noise sources i_1 and i_2, the impedance representation two voltage noise sources u_1 and u_2 and the chain representation a voltage noise source u and a current noise source i.

A physically significant description of these sources is given by their self- and cross-power spectral densities which are defined as the Fourier transform of their auto- and cross-correlation functions.[1] Arranging these spectral densities in matrix form leads to the so-called correlation matrices [6]. The correlation matrices belonging to admittance, impedance, and chain representation are shown in Fig. 1. The elements of matrices are denoted by $C_{s_1 s_2^*}$ where the subscript indicates that the spectral density refers to the noise sources s_1 and s_2. The matrices themselves are denoted by C and by a subscript which specifies the representation. The noiseless part of the noise equivalent two-port is described by electrical matrices. These matrices are the conventional two-port matrices. They are also shown in Fig. 1.

Noise sources are usually characterized by their mean fluctuations in bandwidth Δf centered on frequency f. For two noise sources s_1 and s_2, the mean fluctuations are $\langle s_1 s_1^* \rangle$, $\langle s_1 s_2^* \rangle$, $\langle s_2 s_1^* \rangle$, and $\langle s_2 s_2^* \rangle$ ($\langle s_i s_j^* \rangle$ denotes the mean fluctuation of a product containing the signal s_i and the complex conjugate of the signal s_j). Mean fluctuations are closely related to power spectral densities. This relation has the form

$$\langle s_i s_j^* \rangle = 2\Delta f C_{s_i s_j^*}, \qquad i,j = 1,2. \tag{1}$$

The factor 2 occurs because the frequency range has been taken from $-\infty$ to $+\infty$. The correlation matrix C belonging to the noise sources s_1 and s_2 can then be written as

$$C = \frac{1}{2\Delta f} \begin{bmatrix} \langle s_1 s_1^* \rangle & \langle s_1 s_2^* \rangle \\ \langle s_2 s_1^* \rangle & \langle s_2 s_2^* \rangle \end{bmatrix}. \tag{2}$$

[1] The noise sources are assumed to be stationary random processes.

III. CHANGES OF REPRESENTATION, INTERCONNECTION OF NOISY TWO-PORTS

If two or more representations exist (and they generally do) these representations can be transformed into each other by simple transformation operations. The derivation of these formulas is straightforward. As the system is linear the noise signals of the new representation (denoted by the vector $x'(t)$) can be expressed in terms of the noise signals of the old representation (denoted by the vector $x(t)$) by the convolution integral

$$x'(t) = \int_{-\infty}^{+\infty} H(s) x(t-s) ds \tag{3}$$

where the transformation is characterized by the weighting matrix $H(s)$ [7]. Using this relation the auto- and cross-correlation functions are calculated. Fourier transforming them leads to the transformation formula

$$C' = TCT^+ \tag{4}$$

where C and C' denote the correlation matrix of the original and resulting representation, respectively. The transformation matrix T is the Fourier transform of $H(s)$. The plus sign (+) is used to denote Hermitian conjugation. Obviously, the T matrix depends only upon the system but not upon the noise parameters of the two-port.

A simple procedure for determining the transformation matrix is as follows. The noise sources of both the original and the resulting equivalent circuits are replaced by non-random sinusoidal sources. Now a frequency domain analysis can be applied to establish relations between the Fourier amplitudes of the original and resulting circuit. By expressing these relations in matrix form the desired transformation matrix is obtained.

A set of matrices covering all possible transformations between impedance, admittance, and chain representation is presented in Table I.

Interconnection of noisy two-ports is also formulated by means of operations with correlation matrices. Formulas corresponding to the various types of interconnections can be obtained by the general concept demonstrated above.

In deriving these formulas it is assumed that there is no correlation between the noise sources of different two-ports. This is by no means a restriction to the applicability as far as the basic two-ports correspond to individual devices. In this case, clearly no noise correlation exists. Problems may arise, however, in applications for device modeling. Then the correlation matrix concept has to be extended to a more general form.

For applications in noise analysis interconnections of two two-ports either in parallel, in series or in cascade are of particular interest. For these interconnections the resulting correlation matrix is related to the correlation matrices of the original two-ports by

$$C_Y = C_{Y1} + C_{Y2} \quad \text{(parallel)} \quad (5)$$
$$C_Z = C_{Z1} + C_{Z2} \quad \text{(series)} \quad (6)$$
$$C_A = A_1 C_{A2} A_1^+ + C_{A1} \quad \text{(cascade)} \quad (7)$$

where the subscripts 1 and 2 refer to the two-ports to be connected. As shown by these equations interconnection in parallel and in series corresponds to addition of the correlation matrices in admittance and impedance representation, respectively. For cascading (in an order indicated by the subscripts) a more complicated relation is obtained which additionally contains the electrical matrix A_1 of the first two-port.

IV. Correlation Matrix of the Basic Two-Ports

The two-port analysis starts from basic two-ports whose correlation matrices have to be known. These matrices are obtained from either theoretical considerations or noise measurements. An important example belonging to the former case is the thermal noise of two-ports consisting only of passive elements. On thermodynamic grounds, the correlation matrices in impedance and admittance representation of such a two-port are

$$C_Z = 2kT \, \text{Re} \, \{Z\} \quad (8)$$
$$C_Y = 2kT \, \text{Re} \, \{Y\}. \quad (9)$$

They are completely determined by the temperature T and the real part of their electrical matrices in impedance and admittance representation, respectively. Theoretical estimations of the correlation matrix are also obtained if noise equivalent circuits of the elements are used. This is demonstrated by a simplified transistor model as shown in Fig. 2. The equivalent circuit is considered as a cascade of two two-ports. The correlation matrix of the first two-port is obtained from (9). The correlation matrix of the second two-port is according to (2)

$$C_Y = \frac{1}{2\Delta f} \begin{bmatrix} \langle i_b i_b^* \rangle & \langle i_b i_c^* \rangle \\ \langle i_c i_b^* \rangle & \langle i_c i_c^* \rangle \end{bmatrix}. \quad (10)$$

The quantities $\langle i_b i_b^* \rangle$, $\langle i_b i_c^* \rangle$, $\langle i_c i_b^* \rangle$, and $\langle i_c i_c^* \rangle$ can be taken from the literature, e.g., [1], [2].

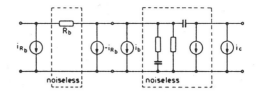

Fig. 2. Simplified noise equivalent circuit for transistor.

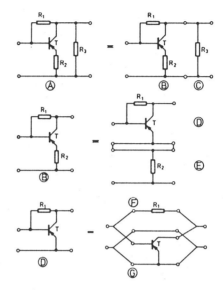

Fig. 3. Principles of noise analysis demonstrated for amplifier.

In cases where the correlation matrix cannot be derived from theory measurements of the noise performance provide the required information. Such measurements are usually done by determining the equivalent noise resistance R_n, the optimal source admittance Y_{opt} and the minimum noise figure NF_{min}. With these quantities estimated the chain representation of the correlation matrix is obtained as

$$C_A = 2kT \begin{bmatrix} R_n & \dfrac{NF_{\text{min}} - 1}{2} - R_n Y_{\text{opt}} \\ \dfrac{NF_{\text{min}} - 1}{2} - R_n Y_{\text{opt}}^* & R_n |Y_{\text{opt}}|^2 \end{bmatrix} \quad (11)$$

where T is the absolute temperature.

V. The Concept of Noise Analysis

The principles of the analysis procedure are explained with reference to Fig. 3 where the amplifier A is to be analyzed. In a first step, a decomposition of the two-port A into basic two-ports C, E, F, and G is carried out. The basic two-ports have to be specified by their electrical and correlation matrices. The electrical matrices are obtained by any usual procedure, either by calculation or by measurement. The correlation matrices are determined using the results of Section IV. Once all matrices are known the basic two-ports are successively interconnected in a manner that finally the two-port A is obtained. For example, two-port D results from interconnecting F and E in parallel. The matrices characterizing two-port D are determined by the following two-step procedure: 1) the matrices of E and F are transformed into admittance representation which is the appropriate representation for parallel interconnection; 2) adding the electrical matrices of D and E yields the electrical matrix of D, adding the correlation matrices the correlation matrix of D (cf. (5)).

It is demonstrated by this example that the correlation matrix of a two-port can be calculated from the matrices of the basic two-ports by a consequent application of the interconnection rules given in Section III. Once the correlation matrix has been determined in chain representation the noise parameters can be computed. Denoting this correlation matrix by C_A and its elements by (C_{uu^*}, C_{ui^*}, C_{ui}^*, C_{ii^*}) the noise figure relating to a source impedance Z_S is given by

$$NF = 1 + \frac{z^+ C_A z}{2kT \, \text{Re} \{Z_S\}} \quad (12)$$

where

$$z = \begin{bmatrix} 1 \\ Z_S^* \end{bmatrix}. \quad (13)$$

This result follows immediately from the definition of the noise figure. Equation (11) is used to express the optimal source admittance and the minimum noise figure as function of the correlation matrix. The following relations are obtained:

$$Y_{opt} = \sqrt{\frac{C_{ii^*}}{C_{uu^*}} - (\text{Im}\{C_{ui^*}/C_{uu^*}\})^2} - j \, \text{Im}\{C_{ui^*}/C_{uu^*}\} \quad (14)$$

$$NF_{min} = 1 + (C_{ui^*} + C_{uu^*} Y_{opt})/kT. \quad (15)$$

Minimum noise figure and optimal source admittance can be directly determined by this noise analysis method.

Basic to this concept is that the analysis of any network consists in a consequent application of only a few operations (change of representation, interconnection, noise figure calculation). This makes it well suited for a computer aided analysis where these operations are left to the computer. The sequence of operations is controlled by a supervisor which can be either the designer itself or the computer. Obviously, there is a large variety of actual implementations. In principle, any of the existing implementations of ac two-port analysis can be extended to include noise analysis.

VI. Conclusions

The noise analysis concept presented in this paper is an application of the circuit theory of linear noisy networks. The correlation matrices represent a systematic description of noise in linear networks which includes both the uncorrelated and the (partially or fully) correlated case. In contrast to other noise analysis methods, partially correlated noise sources can be treated with the same efficiency.

The noise analysis method is based on the two-port analysis concept and is, therefore, suited for a wide class of networks which covers most applications occurring in practice. It can easily be adapted to already existing two-port analysis programs for ac analysis.

The set of noise parameters which can be calculated by the method includes the noise figure, the minimum noise figure, and the optimal source admittance. All noise parameters are obtained by one single analysis.

References

[1] W. Baechtold, W. Kotyczka, and M. J. O. Strutt, "Computerized calculation of small signal and noise properties of microwave transistors," *IEEE Trans. Microwave Theory Technol.*, vol. MTT-17, pp. 614–619, Aug. 1969.
[2] R. Rohrer, L. Nagel, R. Meyer, and L. Weber, "Computationally efficient electronic-circuit noise calculations," *IEEE Trans. Solid-State Circuits*, vol. SC-6, pp. 204–213, Aug. 1971.
[3] F. H. Branin, "Network sensitivity and noise analysis simplified," *IEEE Trans. Circuit Theory*, vol. CT-20, pp. 285–288, May 1973.
[4] K. Hartmann, W. Kotyczka, and M. J. O. Strutt, "Computerized determination of electrical network noise due to correlated and uncorrelated noise sources," *IEEE Trans. Circuit Theory*, vol. CT-20, pp. 321–322, May 1973.
[5] H. A. Haus and R. B. Adler, *Circuit Theory of Linear Noisy Networks*. New York: Wiley, 1959.
[6] H. W. König and H. Pötzl, *Schwankungsvorgänge in Elektronenstrahlen, Fortschritte der Hochfrequenztechnik*, vol. 4, Akademische Verlagsgesellschaft, Frankfurt, Germany, 1959, pp. 195–239.
[7] W. B. Davenport and W. L. Root, *An Introduction to the Theory of Random Signals and Noise*. New York: McGraw-Hill, 1958.

Part IV: Noise In Active Filters
Noise Performance of Low-Sensitivity Active Filters

L. T. BRUTON, F. N. TROFIMENKOFF, AND DAVID H. TRELEAVEN

Abstract—The noise limitations of some important low-sensitivity RC-active filter realizations are derived; the inherent rms noise (noise from essential network resistors) that exists at the output terminals of a bandpass active filter is determined for two-integrator loop and positive impedance converter filter realizations. Output noise spectral density and total rms output noise are determined for the two-integrator loop section. The noise contributed by the essential circuit resistors and the operational amplifiers is taken into account and it is shown that exact calculation of output noise may be obtained in terms of Q factor and impedance level. The results are used to obtain the theoretical noise limitations of these low-sensitivity active filters. It is found that the deterioration in signal-to-noise ratio (SNR), due to the internal filter noise, is independent of Q factor for high Q realizations.

NOTATION

$T(s)$	Voltage transfer function.
ω_0	Resonant frequency, rad/s.
Q	Q factor.
$e_{on}(t), \epsilon_{on}(\omega), E_{on}$	Output noise waveform, spectral density, and rms value, respectively.
$e_{OR}(t), \epsilon_{OR}(\omega), E_{OR}$	Corresponding components of output noise due to inherent resistors.
$e_{OA}(t), \epsilon_{OA}(\omega), E_{OA}$	Corresponding components of output noise due to amplifiers.
$H_{io}(j\omega)$	Voltage transfer function from a voltage generator in series with resistor R_i to the filter output terminals.
$\epsilon_{ni}(\omega)$	Johnson noise spectral density for resistor R_i.
k	Boltzmann's constant.
T	Absolute temperature, Kelvin.
$e_{NAi}(\omega)$	Effective input noise voltage spectral density of amplifier i.
R	Resistance of essential inherent resistors.
R'	Resistance of nonessential resistors.
f_N	Effective noise bandwidth of transfer function $T(j\omega)$, hertz.
$\omega_x \equiv 2\pi f_x$	Upper frequency limit of the noise spectrum.
$N_0(f)$	Spectral density of input noise in the region, $0 < f < f_x$.
I	Improvement in signal to noise ratio (SNR) due to filtering.

Manuscript received March 9, 1972; revised August 21, 1972.
L. T. Bruton and F. N. Trofimenkoff are with the Department of Electrical Engineering, University of Calgary, Calgary, Alta., Canada.
D. H. Treleaven is with Microsystems International, Inc., Ottawa, Ont., Canada.

Note: Spectral densities $\epsilon(\omega)$ have dimensions volts/$\sqrt{\text{radians/second}}$ unless written $\epsilon(f)$ in which case the dimensions volts/$\sqrt{\text{hertz}}$ are implied.

I. INTRODUCTION

IN THE sense that RC-active filters contribute electrical noise to the filtered signal, they are not ideal. This noise will be due to Johnson noise in the resistors and to noise contributed by the active devices. The noise generated by the active devices will be dependent on the particular device that is employed and may consist of thermal noise, flicker noise, and burst noise. In addition, the actual measurable rms noise at the output of a filter will depend on the exact transfer function and topology of the realization. The fact that there are many RC-active filter synthesis techniques, makes a preliminary investigation of active filter noise a difficult task. The noise performance of the low-sensitivity two-integrator loop [1]–[3] and the positive impedance converter/inverter (PIC) [4]–[10] type realizations are analyzed in this work. All of these techniques have led to the realization of high-quality high-Q transfer functions for which a detailed noise performance theory would be useful but is thus far unavailable.

The noise performance of the basic nullor [11], [12] model of the two-integrator loop biquadratic section is derived in terms of the relevant transfer function parameters. The noise calculations obtained from the nullor realization represent the ideal minimum noise level that is available for the network because the nullor corresponds to an ideal noise-free infinite gain controlled source; for this reason, the noise due to the nullor network is defined as the inherent noise. Since the two-integrator loop biquadratic section is the nullor equivalent of a class of two-amplifier PIC's, it can be shown that the inherent-noise performance of these two types of filter is identical. It is found that inherent noise is a function of the Q factor, the center frequency ω_0, and of the Johnson thermal noise contributed by the network resistors.

The additional noise contributed by the noisy active devices, which must be used to replace the nullors, is defined as "amplifier noise" and is derived in detail for the two-integrator loop. This particular realization has been chosen for the more detailed analysis of amplifier noise for several reasons.

1) The transfer functions from each effective noise generator to the output of the filter are relatively easy to determine; hence, the total noise performance may be determined.

2) The two-integrator realization is now widely used

as a building block in the realization of microelectronic filters and in analog computing.

3) The dynamic range and noise performance limitations of this circuit had not been investigated previously.

II. Two-Integrator Biquadratic Section

The two integrator biquadratic section is a particularly useful low-sensitivity realization [2] that may be used as the basic building block for the realization of low-, band and high-pass voltage transfer functions. Variations of the basic realization in Fig. 1 have been proposed [1] and analyzed [3]. This network has a bandpass voltage transfer function given by

$$T(s) \equiv \frac{-1}{\omega_0 C_3 R_1} \cdot \left[\frac{(s/\omega_0)}{(s/\omega_0)^2 + (s/Q\omega_0) + 1} \right], \quad (1)$$

where the resonant frequency ω_0 is given by

$$\omega_0^2 = R_5/C_3 C_7 R_2 R_4 R_6 \quad (2)$$

and the Q factor by

$$Q = \omega_0 C_3 R_3. \quad (3)$$

The transfer function $|T(j\omega)|$ has been shown to exhibit a low-valued multiparameter sensitivity function [13] and this type of realization may be used to obtain very high Q factors. Thus, this type of high-Q bandpass realization is a most attractive method of achieving ultranarrow-band filtering. This work is primarily concerned with the noise performance and associated dynamic range limitations of this type of high-quality bandpass realization.

The noise voltage $e_{on}(t)$ at the output port of Fig. 1 will be considered in terms of its two component parts: the inherent resistor output noise $e_{OR}(t)$ and the active device output noise $e_{OA}(t)$, where we define these separate components of the total noise as follows: $e_{OR}(t) \equiv$ output port noise voltage due to resistors $R_1 \cdots R_6$ and $e_{OA}(t) \equiv$ output noise voltage due to amplifier noise sources and offset compensating resistors R_8–R_{10}.

It should be clear that $e_{OR}(t)$ is the inherent noise voltage that would be measured at the output if it were possible to employ infinite-gain noise-free amplifiers with no offset resistors; $e_{OR}(t)$ depends on the Johnson noise in the resistors $R_1 \cdots R_6$ and the transfer functions $H_{10}(j\omega), H_{20}(j\omega) \cdots H_{60}(j\omega)$ from each of these resistors to the output port. The amplifier-noise voltage $e_{OA}(t)$ is an additional component that is always observed in practice and must be determined from a suitable noise model of the noisy amplifiers and resistors R_8, R_9, R_{10}.

III. Inherent Noise

The inherent noise output $e_{OR}(t)$ may be obtained by replacing each noisy amplifier by a nullor, that is, by an ideal noise-free controlled source. The circuit noise is then the minimum possible and is due only to the essential resistors $R_1 \cdots R_6$.

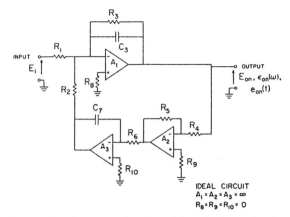

Fig. 1. Two-integrator bandpass biquadratic realization.

The inherent noise model of the two-integrator biquadratic section is given in Fig. 2 where the noise generators $e_{n1}(t) \cdots e_{n6}(t)$ represent the Johnson noise voltage of the corresponding resistors $R_1 \cdots R_6$. The transfer functions from each one of $e_{n1}(t) \cdots e_{n6}(t)$ to the output port are defined by $H_{10}(j\omega) \cdots H_{60}(j\omega)$ and are given in Table I where $H_{LP}(j\omega)$ and $H_{BP}(j\omega)$ are normalized low- and bandpass transfer functions defined as follows:

$$H_{LP}(j\omega) \equiv \frac{1}{1 - \left(\frac{\omega}{\omega_0}\right)^2 + j\frac{\omega}{\omega_0 Q}} \quad (4)$$

and

$$H_{BP}(j\omega) \equiv j\frac{\omega}{\omega_0} \cdot H_{LP}(j\omega). \quad (5)$$

The thermal noise voltage generator $e_{ni}(t)$ for the resistor R_i may be represented [14] by a spectral density noise function $\epsilon_{ni}(\omega)$, where $\epsilon_{ni}^2(\omega) = (2kT/\pi)R_i$ so that the spectral density $\epsilon_{OR}(\omega)$ of the inherent output noise $e_{OR}(t)$ is given by

$$\epsilon_{OR}^2(\omega) = \frac{2kT}{\pi} \sum_{i=1}^{6} R_i |H_{i0}(j\omega)|^2 \quad (6)$$

and the total inherent rms output noise E_{OR} by

$$E_{OR}^2 = \int_0^\infty \epsilon_{OR}^2(\omega) \cdot d\omega$$

or

$$E_{OR} = \sqrt{\frac{2kT}{\pi} \int_0^\infty \sum_{i=1}^{6} R_i |H_{i0}(j\omega)|^2 \cdot d\omega}. \quad (7)$$

This expression for the inherent rms output noise is fortunately solvable in closed form. Comparing (7) and column one of Table I, it is apparent that only two integrals are required to evaluate E_{OR}, which are

$$\int_0^\infty |H_{LP}(j\omega)|^2 \cdot d\omega$$

and

$$\int_0^\infty |H_{BP}(j\omega)|^2 \cdot d\omega.$$

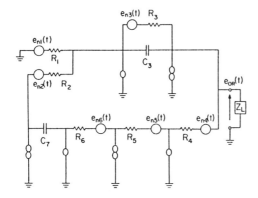

Fig. 2. Inherent noise model of the two-integrator biquadratic realization.

TABLE I

$|H_{io}(j\omega)|^2$	$\int_0^\infty |H_{io}(j\omega)|^2 \, d\omega$
$|H_{10}(j\omega)|^2 = \left(\dfrac{1}{\omega_0 R_1 C_3}\right)^2 |H_{BP}(j\omega)|^2$	$\left(\dfrac{1}{R_1 C_3}\right)^2 \dfrac{Q\pi}{2\omega_0}$
$|H_{20}(j\omega)|^2 = \left(\dfrac{1}{\omega_0 R_2 C_3}\right)^2 |H_{BP}(j\omega)|^2$	$\left(\dfrac{1}{R_2 C_3}\right)^2 \dfrac{Q\pi}{2\omega_0}$
$|H_{30}(j\omega)|^2 = \left(\dfrac{1}{\omega_0 R_3 C_3}\right)^2 |H_{BP}(j\omega)|^2$	$\left(\dfrac{1}{R_3 C_3}\right)^2 \dfrac{Q\pi}{2\omega_0}$
$|H_{40}(j\omega)|^2 = |H_{LP}(j\omega)|^2$	$\dfrac{Q\pi\omega_0}{2}$
$|H_{50}(j\omega)|^2 = \left(\dfrac{R_4}{R_5}\right)^2 |H_{LP}(j\omega)|^2$	$\left(\dfrac{R_4}{R_5}\right)^2 \dfrac{Q\pi\omega_0}{2}$
$|H_{60}(j\omega)|^2 = \left(\dfrac{R_4}{R_5}\right)^2 |H_{LP}(j\omega)|^2$	$\left(\dfrac{R_4}{R_5}\right)^2 \dfrac{Q\pi\omega_0}{2}$

The evaluation of these integrals gives

$$\int_0^\infty |H_{LP}(j\omega)|^2 \, d\omega = \int_0^\infty |H_{BP}(j\omega)|^2 \, d\omega = \frac{Q\pi\omega_0}{2}.$$

Thus, the application of this result allows the second column of Table I to be completed; then substituting the results given in Table I into (7) gives

$$E_{OR}^2 = QkT\left[\frac{1}{R_1 C_3^2 \omega_0} + \frac{1}{R_2 C_3^2 \omega_0} + \frac{1}{R_3 C_3^2 \omega_0} + \frac{R_4^2 \omega_0}{R_5} + R_4\omega_0 + \left(\frac{R_4}{R_5}\right)^2 R_6\omega_0\right]. \quad (8)$$

The result in (8) is exact and may always be used to calculate the inherent rms output noise due to any combination of $C_3, C_7, R_1 \cdots R_6$. However, it is possible to obtain physical interpretation of this result without loss of generality. For many practical designs, it is convenient to select $R_2 = R_4 = R_5 = R_6 \equiv R$ and C_3 $= C_7 \equiv C$ so that, from (2) and (3),

$$\omega_0 = 1/CR \qquad Q = R_3/R. \quad (9)$$

Then, substituting (9) into (8) gives

$$E_{OR} = \sqrt{kTR(5Q+1)\omega_0} \quad (10)$$

It should be noted that this result is exact; it does not depend on $Q \gg 1$ for validity.

Having evaluated the total inherent rms output noise, it is worthwhile to comment on the inherent output spectral density function $\epsilon_{OR}(\omega)$ as given by (6). For $Q \gg 1$, the $|H_{LP}(j\omega)|$ and $|H_{BP}(j\omega)|$ functions are almost identical in the region of ω close to ω_0. Thus, for $Q \gg 1$ and ω close to ω_0,

$$\epsilon_{OR}(\omega) \approx \sqrt{(10kTR/\pi) \cdot H_{BP}(j\omega)} \quad (11)$$

and consequently the spectral density function $\epsilon_{OR}(\omega)$ is approximately the same shape as the filter transfer function $|T(j\omega)|$ and has a peak value given by

$$\epsilon_{OR}(\omega)_{\text{peak}} \approx \sqrt{(10kTR/\omega) \cdot Q} \quad (12)$$

Typical Example of Inherent Noise

Consider a bandpass design with the following design parameters,

$$Q = 150 \quad (13)$$
$$f_0 = 1500 \text{ Hz} \quad (14)$$

given by

$$R = 10 \text{ k}\Omega \qquad kT/q = 0.026 \text{ V at } 300 \text{ K}$$
$$R_3 = 1.5 \text{ M}\Omega \quad (15)$$
$$C = 0.106 \text{ }\mu\text{F} \qquad q = 1.6 \times 10^{-9} \text{ C}.$$

Then, direct substitution into (10) and (12) gives

$$E_{OR} = 17.3 \text{ }\mu\text{V rms} \quad (16)$$
$$\epsilon_{OR}(\omega)_{\text{peak}} = 4.37 \text{ }\mu\text{V}/\sqrt{\text{Hz}}. \quad (17)$$

IV. AMPLIFIER NOISE

In the previous section, the minimum possible output noise has been determined. The purpose of this section is to determine the additional noise contribution that is due to the noise sources that are inevitably associated with any practical realization of the nullors. For the purpose of this contribution, it is assumed that the realization is that given in Fig. 1; that is, noisy high-gain operational amplifiers correspond to the nullors and the provision of dc offset (noisy) resistors R_8, R_9, and R_{10} are taken into account. This latter consideration is in fact necessary because, as will be shown, the dc offset resistors contribute significantly to the total circuit noise output.

It is known [15], [16] that a noisy operational amplifier may be represented by three effective input noise generators, which may be two effective noise current generators i_{n1} and i_{n2} and one effective noise voltage generator e_{nA}, as shown in Fig. 3(a). The operational amplifier in Fig. 3(a) may be regarded as noise free and the effective noise generators e_{nA}, i_{n1}, and i_{n2} may be determined for any particular operational amplifier by direct measurement or from the manufacturer's specification. The general active device amplifier noise model of a single stage of Fig. 1 is given in Fig. 3(a), where e_{nR} is due to the Johnson noise of the dc offset resistance R. A direct analysis of the total noise output spectral density $\epsilon_{OA}(\omega)$ due to the amplifier noise models is somewhat complicated and lengthy. However, it is easily shown that the amplifier noise model of Fig. 3(a) is electrically equivalent to that given in Fig. 3(b), where the noise generators now appear in series with the input and output terminals of a complete noise-free amplifier stage containing Z_1, Z_2, R, and the ideal operational amplifier. This equivalent amplifier noise model allows the output spectral density $\epsilon_{OA}(\omega)$ to be calculated using the previously determined transfer functions $|H_{io}(j\omega)|^2$, because the noise generators are now effectively in series with the resistors $R_1 \cdots R_6$.

To simplify the analysis, it is possible to make a reasonable practical assumption; that is, in terms of Fig. 3(b), $i_{n2}(\omega)Z_1(\omega) \ll e_{nA}(\omega)$ and $i_{n1}(\omega)Z_2(\omega) \ll e_{nA}(\omega)$. (Typically, i_{n1}, i_{n2} are about 0.1 pA/$\sqrt{Hz}$ and e_{nA} is about 10 nV/$\sqrt{Hz}$. Consequently, the above assumption is valid for $|Z_1(\omega)|$ and $|Z_2(\omega)|$ equal to or less than about 100 kΩ.) With this assumption, the amplifier noise model is given by the circuit in Fig. 4. The output noise spectral density $\epsilon_{OA}(\omega)$ may now be evaluated directly from the transfer functions in Table I. Direct analysis of Fig. 4, using Table I, gives the output spectral density $\epsilon_{OA}(\omega)$ as

$$\epsilon_{OA}^2(\omega) = [\epsilon_{NA1}^2 + \epsilon_{n8}^2 + \epsilon_{NA3}^2 + \epsilon_{n10}^2] \cdot |H_{BP}(j\omega)|^2$$
$$+ \left[\epsilon_{NA2}^2 + \epsilon_{n9}^2 + (\epsilon_{NA2}^2 + \epsilon_{n9}^2 + \epsilon_{NA3}^2 + \epsilon_{n10}^2)\left(\frac{R_4}{R_5}\right)^2\right] \cdot |H_{LP}(j\omega)|^2$$
$$+ [(\epsilon_{NA1}^2 + \epsilon_{n8}^2)] \cdot |H_{HP}(j\omega)|^2, \quad (18)$$

where $H_{HP}(j\omega)$ is the normalized second-order high pass transfer function defined by

$$H_{HP}(j\omega) \equiv \frac{j\omega}{\omega_0} \cdot H_{BP}(j\omega). \quad (19)$$

Without loss of generality, (18) may be simplified by once more assuming that (9) is valid, giving

$$\epsilon_{OA}^2(\omega) = [(\epsilon_{NA1}^2 + \epsilon_{n8}^2) + (\epsilon_{NA3}^2 + \epsilon_{n10}^2)] \cdot |H_{BP}(j\omega)|^2$$
$$+ [2(\epsilon_{NA2}^2 + \epsilon_{n9}^2) + (\epsilon_{NA3}^2 + \epsilon_{n10}^2)] \cdot |H_{LP}(j\omega)|^2$$
$$+ [(\epsilon_{NA1}^2 + \epsilon_{n8}^2)] \cdot |H_{HP}(j\omega)|^2. \quad (20)$$

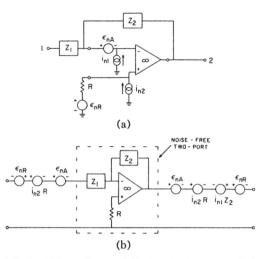

Fig. 3. (a) Amplifier-noise model of a single stage of the network. (b) Equivalent amplifier-noise model of a single stage of the network.

Now, $|H_{HP}(j\omega)|^2$ may be eliminated from the previous equation by noting, from (4), (5), and (19) that

$$|H_{HP}(j\omega)|^2 = [2 - (1/Q^2)]|H_{BP}(j\omega)|^2$$
$$- |H_{LP}(j\omega)|^2 + 1 \quad (21)$$

so that, substituting (21) into (20) gives

$$\epsilon_{OA}^2(\omega) = [2(\epsilon_{NA1}^2 + \epsilon_{n8}^2) + (\epsilon_{NA3}^2 + \epsilon_{n10}^2)] \cdot |H_{BP}(j\omega)|^2$$
$$+ [2(\epsilon_{NA2}^2 + \epsilon_{n9}^2) + (\epsilon_{NA3}^2 + \epsilon_{n10}^2)$$
$$- (\epsilon_{NA1}^2 + \epsilon_{n8}^2)] \cdot |H_{LP}(j\omega)|^2$$
$$+ [\epsilon_{NA1}^2 + \epsilon_{n8}^2]. \quad (22)$$

The previous equation is used to evaluate the output amplifier noise spectral density $\epsilon_{OA}(\omega)$. The total rms output amplifier noise E_{OA} is given by

$$E_{OA}^2 = \int_0^{\omega_x} \epsilon_{OA}^2(\omega) \cdot d\omega, \quad (23)$$

where ω_x is the maximum bandwidth of the noise measurement. Then, performing this integration on (22), for $\omega_x \gg \omega_0$, gives

$$E_{OA}^2 \approx [(\epsilon_{NA1}^2 + \epsilon_{n8}^2) + 2(\epsilon_{NA3}^2 + \epsilon_{n10}^2)$$
$$+ 2(\epsilon_{NA2}^2 + \epsilon_{n9}^2)](Q\pi\omega_0/2) + \int_0^{\omega_x}(\epsilon_{NA1}^2 + \epsilon_{n8}^2) \cdot d\omega$$

and substituting the Johnson noise spectral densities for ϵ_{n8}, ϵ_{n9}, and ϵ_{n10} (that is, $e_{ni}^2 = 2kTR_i/\pi$ V^2/rad/s) gives

$$E_{OA}^2 \approx \underbrace{\frac{Q\pi\omega_0}{2}[\epsilon_{NA1}^2 + 2\epsilon_{NA2}^2 + 2\epsilon_{NA3}^2] + \int_0^{\omega_x}\epsilon_{NA1}^2 d\omega}_{\text{due to amplifiers}}$$
$$+ \underbrace{\frac{Q\pi\omega_0}{2}\left[\frac{4kT}{\pi}(R_9 + R_{10}) + \frac{2kT}{\pi}R_8\right] + \frac{2kT}{\pi}R_8\omega_x}_{\text{due to offset resistances}}.$$

The previous equation can be considerably simplified by assuming that $e_{NA1} = e_{NA2} = e_{NA3} \equiv e_{NA}$ and $R_8 = R_9 = R_{10} \equiv R'$, giving

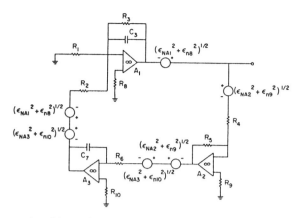

Fig. 4. Amplifier-noise model of the two-integrator loop biquadratic realization.

$$E_{OA}^2 \approx \frac{5Q\pi\omega_0\epsilon_{NA}^2}{2} + 5Q\omega_0 kTR'$$
$$+ \left[\int_0^{\omega_x} \epsilon_{NA}^2 \cdot d\omega + \frac{2\omega_x kTR'}{\pi}\right].$$

The amplifier noise E_{OA} is a function of the measurement bandwidth ω_x and is due to the high pass function $H_{HP}(j\omega)$ that exists between ϵ_{NA1} and ϵ_{n8} and the output of the filter. If the Q factor is large, it is usually possible to assume that

$$Q \gg \omega_x/\omega_0 \tag{25}$$

so that (24) becomes

$$E_{OA}^2 \approx 5Q^2 f_N [\epsilon_{NA}^2(f) + 4kTR'], \tag{26}$$

where $f_N \equiv \pi/2 \cdot f_0/Q$, the effective noise bandwidth of the high-Q bandpass filter.

Typical Example of Amplifier Noise

Consider the bandpass filter that was specified by (13)–(15) and assume that the offset resistors are given by

$$R' = 10 \text{ k}\Omega \tag{27}$$

and that the noise voltages of the amplifiers are given by

$$\epsilon_{NA}(f) = 15.1 \text{ nV}/\sqrt{\text{Hz}}. \tag{28}$$

The previous assumed value of $\epsilon_{NA}(f)$ corresponds to the measured data for a sample of Motorola 1741CL operational amplifiers that have been used for the practical verification of these results. Substituting (13), (14), (27), and (28) into (26) gives

$$E_{OA} \approx 26.5 \text{ }\mu\text{V rms}. \tag{29}$$

Total Output Noise E_{on}

To evaluate the total rms output noise E_{on} it is only necessary to add the uncorrelated rms noise components E_{OR} and E_{OA}. Now, assuming that $Q \gg 1$ so that, from (10) yields

$$E_{OR}^2 \approx 5Q\omega_0 kTR = 5Q^2 f_N [4kTR], \tag{30}$$

so that

$$E_{on}^2 = E_{OA}^2 + E_{OR}^2$$
$$= 5Q^2 f_N [4kTR + 4kTR' + \epsilon_{NA}^2(f)]$$

or

$$E_{on} = Q\sqrt{5f_N[4kTR + 4kTR' + \epsilon_{NA}(f)]}.$$

or

$$E_{on} = Q\sqrt{5f_N[\epsilon_{nR}^2(f) + \epsilon_{nR'}(f) + \epsilon_{NA}^2(f)]}, \tag{31}$$

where $\epsilon_{nR}(f)$ and $\epsilon_{nR'}(f)$ are the Johnson noise spectral densities of the resistors R and R'. For the previously considered filter parameters

$$E_{on} = \sqrt{17.3^2 + 26.5^2} \text{ } \mu\text{V rms} = 31.7 \text{ } \mu\text{V rms}. \tag{32}$$

V. EXPERIMENTAL VERIFICATION OF FILTER OUTPUT NOISE PERFORMANCE

The circuit in Fig. 1 was constructed in the laboratory; $R_1, R_2, R_3 \cdots R_{10}$ were selected to be 10 kΩ and the capacitors $C_7 = C_3 = C$ to give a resonant frequency f_0 of 1500 Hz. Then, R_3 was adjusted to give a Q factor of 150 so that all parameters correspond as closely as possible to the numerical values assumed in the calculation of E_{on} (32). The Motorola 1741CL operational amplifiers were employed so that the measured values of $\epsilon_{NA}(f)$ (28) were known to be valid. It was found that the spread in $\epsilon_{NA}(f)$ was less than 10 percent between devices and that $\epsilon_{NA}(f)$ was approximately flat over the range 500–5000 Hz. The measurement of the output noise spectral density $\epsilon_{on}(f)$ was made by measuring the rms noise in a 10-Hz band using the Hewlett-Packard 3590A wave analyzer. The total rms noise E_{on} was obtained by measuring the noise in a 1000-Hz bandwidth. Since the response of the filter transfer functions H_{LP}, H_{BP}, and H_{HP} are negligibly small at ±500 Hz from the 1500-Hz center frequency, it is possible to interpret the 1000-Hz bandwidth measurement as representing E_{on}.

The measured results are given in Fig. 5. The spectral density data are a good approximation to the shape of the voltage transfer function $|T(j\omega)|$ of the filter. This is to be expected because, for this high-Q realization, $|T(j\omega)|$ and $|H_{io}(j\omega)|$ are the same general shape in the region for which a significant filter response is measurable. The total measured output noise is, from Fig. 5, given by

$$E_{on}_{\text{measured}} = 37.8 \text{ } \mu\text{V rms}. \tag{33}$$

Comparison of (32) and (33) indicates good agreement between the measured and theoretical rms output noise. Furthermore, the shape of the noise spectral density curve (compared with $|T(j\omega)|$) confirms the theoretical expressions given by (6) and (22).

VI. OTHER LOW-SENSITIVITY ACTIVE-FILTER REALIZATIONS

The previous derivations of the amplifier noise E_{OA} is applicable only to the two-integrator loop biquadratic

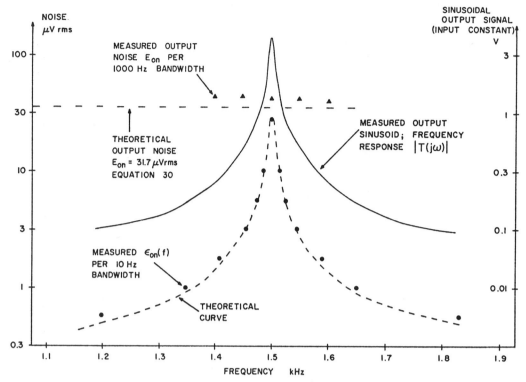

Fig. 5. Measured noise spectra and frequency response.

circuit of Fig. 1. However, the inherent noise E_{OR} derivation is valid for the nullor network in Fig. 2 and consequently is valid for any network that is a nullor equivalent of this network. Now, rearranging the passive elements of Fig. 2 in the form of a hexagon and replacing the three nullors by a two-nullor equivalent [18], [19] leads to the class of positive impedance converters (PIC's) that have been considered elsewhere [4]; thus, the PIC in Fig. 6 is an example of a realization that has an identical inherent noise performance to that of the two-integrator loop because, apart from the omissions of R_1 and R_3, its nullor version is identical to Fig. 2. In fact, this omission reduces the expression for E_{OR} from that given in (30) to

$$E_{\text{OR}} \approx \sqrt{4QkTR} \quad \text{V rms.} \quad (34)$$

Thus, we now have the inherent noise performance of the class of two-amplifier PIC's and consequently a measure of the noise performance to be expected when PIC's of this type are interconnected to realize higher order ladder structures. In fact, Sheahan and Orchard [17] report that their preliminary results of noise levels in PIC ladder filters indicate a contribution of an additional 7 dB (over and above resistor noise) due to amplifiers. This result corresponds to the increase from 17.3 to 31.7 μV rms that has been calculated here for the two-integrator loop; this is in fact equal to a theoretical increase of 5.3 dB due to the amplifiers. Thus, the noise calculations presented here, particularly as they apply to inherent noise, may be used to investigate the noise per-

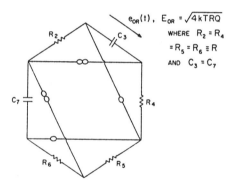

Fig. 6. Positive impedance converter two-nullor equivalent of Fig. 2; R_3 and R_1 omitted.

formance of the wide range of PIC-synthesized filters [1], [3].

VII. Performance Limitations of a Noisy Active Filter

Consider the use of the above bandpass filter configuration for the purpose of improving the signal-to-noise (SNR) of a narrow-band signal contaminated by "flat" band-limited white noise. This is a very common problem in filtering and typically might correspond to recovering a sinusoidal signal from band-limited white noise. The objective of this section is to show that, for a sufficiently low-level input noise spectral density, the improvement in SNR is drastically reduced because the additive filter noise is significant compared to that eliminated by the ideal filtering process.

The input noise spectral density is assumed to be flat, given by N_0 volts/$\sqrt{\text{hertz}}$, over the range 0 to f_x and zero outside this range. Thus, the input noise "power" is given by

$$P_{N_{in}} = N_0^2 f_x.$$

It follows that the filtered noise at the output has a noise power of $Q^2 N_0^2 f_N$. In addition, there is the active-filter noise contribution given by (26). Thus, the total output noise is given by

$$P_{N_{out}} = Q^2 N_0^2 f_N + 5Q^2 f_N (e_{nR}^2(f) + e_{nR'}^2(f) + e_{NA}^2(f))$$

giving

$$\frac{P_{N_{out}}}{P_{N_{in}}} = Q^2 \left[\frac{N_0^2 f_N + 5f_N(e_{nR}^2 + e_{nR'}^2 + e_{NA}^2)}{N_0^2 f_x} \right]. \quad (35)$$

Now, the filter will provide a power gain of Q^2 to a signal at or near the center frequency f_0 and thus from (35), it follows that the improvement in SNR of the noisy filter is given by I, where

$$I = \left[\frac{N_0^2}{N_0^2 + 5(e_{nR}^2 + e_{nR'}^2 + e_{NA}^2)} \right] \frac{f_x}{f_N} \quad (36)$$

Clearly, a noiseless filter gives a SNR improvement of f_x/f_N; for example, if $f_x =$ kHz, then the previously designed filter gives a noiseless filter improvement of

$$I = f_x/f_N = 10^4/\pi \cdot 10 = 318. \quad (37)$$

However, if the input spectral density N_0 is low, then this improvement may be drastically reduced. For example, if

$$N_0 = 5 \text{ nV}/\sqrt{\text{Hz}} \quad (38)$$

and ϵ_{NR}, $\epsilon_{NR'}$, and ϵ_{NA} are given by the previously used values of 13, 13, and 15.1 nV/$\sqrt{\text{Hz}}$, respectively, (36) gives $I = 13.4$. The actual improvement in SNR due to filtering is then much lower than the ideal value given by (37).

Dynamic Range: The determination of the dynamic range of an active filter is of course trivial when the noise level is known. Thus, for the previous practical filter with a theoretical output noise level given by $E_{on} = 31.7$ μV rms, the maximum output sinusoidal signal level was 10 V rms giving a maximum dynamic range of $(10/31.7 \times 10^{-6}) = 3.15 \times 10^5$ or 110 dB.

VIII. Summary

The noise performance of the two-integrator loop bandpass section has been derived in detail; it was found that the output noise spectral density may be obtained in terms of the filter transfer function and the known noise sources within the filter. In particular, the noise contributions due to the inherent resistors, amplifiers, and offset resistors have been evaluated. By assuming a high Q realization, it has been shown that simple expressions for output rms noise are obtained; furthermore, the measured and theoretical output noise are in good agreement. In the practical bandpass circuit under consideration, it was found that the contributions due to noise from amplifiers and offset resistors were a factor of 1.83 (5.3 dB) higher than the inherent noise.

Derived expressions for inherent noise are applicable to other active networks with the same nullor equivalent network; thus, the inherent noise derivations are valid for PIC's and gyrators used in the high-performance filters that have been proposed and constructed by other researchers.

The class of realizations that has been studied here exhibits transfer function magnitude sensitivities that are low valued and independent of Q factor; the calculated improvement in SNR (36) is also independent of Q factor.

Finally, it has been shown that the internal noise generated by an active filter results in a deterioration in the SNR that is significant if the input noise spectral density N_0 is comparable in magnitude to, or less than, the spectral densities ϵ_{NR}, $\epsilon_{NR'}$, or ϵ_{NA}. This deterioration in SNR imposes an inherent limitation in the ability of the active filter to operate effectively at low signal and noise levels.

References

[1] F. E. J. Girling and E. F. Good, "Active filters—The two integrator loop," *Wireless World*, pp. 76–80, Feb. 1970.
[2] W. J. Kerwin, L. P. Huelsman, and R. W. Newcomb, "State-variable synthesis for insensitive circuit transfer functions," *IEEE J. Solid-State Circuits*, vol. SC-2, pp. 87–92, Sept. 1967.
[3] J. Tow, "Design formulas for active RC filters using operational amplifier biquad," *Electron. Lett.*, vol. 5, pp. 339–341, July 1969.
[4] J. Gorski-Popiel, "RC-active synthesis using positive-immittance converters," *Electron. Lett.*, vol. 3, pp. 381–382, Aug. 1967.
[5] R. H. S. Riordan, "Simulated inductance using differential amplifiers," *Electron. Lett.*, vol. 3, pp. 50–51, Feb. 1967.
[6] A. Antoniou, "Realization of gyrators using operational amplifiers and their use in RC-active network synthesis," *Proc. Inst. Elec. Eng.* (London), vol. 116, pp. 1838–1850, Nov. 1969.
[7] L. T. Bruton, "Nonideal performance of two-amplifier positive-impedance converters," *IEEE Trans. Circuit Theory*, vol. CT-17, pp. 541–549, Nov. 1970.
[8] B. A. Shenoi, "Practical realization of a gyrator circuit and RC-gyrator filters," *IEEE Trans. Circuit Theory*, vol. CT-12, pp. 374–380, Sept. 1965.
[9] W. H. Holmes, S. Grutzmann, and W. E. Heinlein, "Direct-coupled gyrators with floating ports," *Electron. Lett.*, vol. 3, pp. 45–46, Feb. 1967.
[10] D. F. Sheahan and H. J. Orchard, "Integratable gyrator using M.O.S. and bipolar transistors," *Electron. Lett.*, vol. 22, p. 390, 1966.
[11] H. J. Carlin, "Singular network elements," *IEEE Trans. Circuit Theory*, vol. CT-11, pp. 67–72, Mar. 1964.
[12] A. C. Davies, "The significance of nullators, norators and nullors in active network theory," *Radio Eng.*, vol. 34, pp. 259–267, 1967.
[13] S. K. Mitra, "Equivalent circuits of gyrators," *Electron. Lett.*, vol. 3, pp. 333–334, July 1967.
[14] H. Nyquist, *Phys. Rev.*, vol. 32, p. 110, 1928.
[15] L. Smith and D. H. Sheingold, "Noise and operational amplifier circuits," in *Analogue Dialogue*, Analogue Devices, Inc., vol. 3, Mar. 1969.
[16] W. R. Huber, "Two-port equivalent noise generators," *Proc. IEEE* (Lett.), vol. 58, pp. 807–809, May 1970.
[17] H. J. Orchard and D. F. Sheahan, "Inductorless bandpass filters," *IEEE J. Solid-State Circuits*, vol. SC-5, pp. 108–118, June 1970.
[18] J. Bendik, "Equivalent gyrator networks with nullators and norators," *IEEE Trans. Circuit Theory* (Corresp.), vol. CT-14, p. 98, Mar. 1967.
[19] L. T. Bruton, "Circuit techniques for use on microelectronic communication and control systems," Ph.D. dissertation, Univ. Newcastle, Newcastle Upon Tyne, England, Nov. 1969.

Noise Performance of *RC*-Active Quadratic Filter Sections

F. N. TROFIMENKOFF, DAVID H. TRELEAVEN, AND L. T. BRUTON

Abstract — A method for calculating the noise voltage at the output of quadratic filter sections is developed. Multiple-feedback low-pass, bandpass, and high-pass quadratic filter sections realized using differential-input single-ended output operational amplifiers are analyzed. The amplifiers are assumed to have infinite input impedance, infinite gain, and zero output impedance. The noise sources associated with the amplifiers are assumed to be statistically independent, but can have both white and $1/f$ noise components. A noise analysis of a fourth-order maximally flat low-pass filter realized by cascading two quadratic filter sections is included.

Manuscript received April 12, 1972; revised November 6, 1972 and January 8, 1973.
F. N. Trofimenkoff and L. T. Bruton are with the Department of Electrical Engineering, University of Calgary, Calgary, Alta., Canada.
D. H. Treleaven is with the Integrated Circuit Development Group, Microsystems International, Ottawa, Ont., Canada.

I. Introduction

DESIGNERS of active filters are concerned with noise performance because both the dynamic range of an active filter and the effectiveness with which a signal can be recovered from a noise background using an active filter are dependent on the output noise of the filter itself. Since $2n$th-order filters can be realized by cascading suitably designed *RC*-active quadratic filter sections, a study of the noise performance of such filter sections is a logical first step in the study of the noise performance of higher order filters of this type.

The output noise of *RC*-active quadratic filter sections is due to Johnson noise [1] in the circuit resistances and to the noise sources associated with the active device or devices used in the

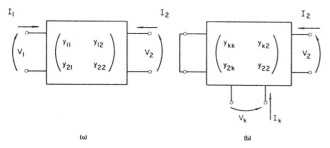

Fig. 1. (a) Conventional two-port network. (b) Technique for dealing with an internal voltage source in a two-port network.

realization of the section. In any particular case, the output noise level will depend on the voltage transfer function of the section, the quality of the active device or devices, the general impedance level of the circuit, and on the topology of the particular realization. Even single-amplifier realizations of quadratic filter sections employing two or three additional network resistors may require as many as five or six noise sources for a complete noise characterization. Thus because a complete noise analysis of an RC-active quadratic filter section can be complicated, considerable emphasis has been placed on the form of the noise analysis and on the method of presenting the results. The noise-analysis techniques are applicable to all two-capacitor quadratic section realizations and to some special types of realizations that employ more than two capacitors. In particular, subject to the above reservation, the techniques are directly applicable to quadratic sections that are realized using Shenoi-type gyrators [2], biquadratic generalized impedance converters [3], Sallen-Key structures [4], single-feedback operational amplifier structures [5], and to structures obtained using the Bach procedure [6].

II. Noise-Analysis Technique

A filter section can be viewed as a two-port network, as shown in Fig. 1(a), and can consequently be described by the equations

$$I_1 = y_{11} V_1 + y_{12} V_2$$
$$I_2 = y_{21} V_1 + y_{22} V_2 \quad (1)$$

where y_{11}, y_{12}, y_{21}, and y_{22} are the short-circuit admittance parameters. The open-circuit output voltage V_2 due to an input voltage V_1 will be given by

$$\frac{V_2}{V_1} = -\frac{y_{21}}{y_{22}}. \quad (2)$$

The total open-circuit output noise voltage of the filter section must be obtained by forming the appropriate sum of the contributions to the open-circuit output noise voltage arising from the various noise sources in the filter section. If the input is short-circuited, as shown in Fig. 1(b), a noise source V_k in a branch of the filter network can be regarded as an input to a new two-port network described by the short-circuit admittance parameters y_{kk}, y_{k2}, y_{2k}, and y_{22}. The open-circuit output voltage V_2 due to the noise voltage V_k will be given by

$$\frac{V_2}{V_k} = -\frac{y_{2k}}{y_{22}}. \quad (3)$$

If the discussion is limited to two-capacitor RC-active networks in which the electrical parameters of the active devices are independent of frequency, all the short-circuit admittance parameters must take the form

$$y_{mn} = \frac{a_{mn} + b_{mn} s + c_{mn} s^2}{a_0 + b_0 s + c_0 s^2} \quad (4)$$

where a_{mn}, b_{mn}, c_{mn}, a_0, b_0, and c_0 are constants and s is the usual Laplace transform variable. It therefore follows that all of the y_{2k}/y_{22} must take the form

$$\frac{a_{2k} + b_{2k} s + c_{2k} s^2}{a_{22} + b_{22} s + c_{22} s^2}. \quad (5)$$

The low-pass (LP), bandpass (BP), and high-pass (HP) quadratic filter sections that are considered in this work have input-output voltage transfer functions of the form

$$T_{LP}(j\omega) = H_{OLP} H_{LP}(j\omega) = \frac{H_{OLP}}{1 - \left(\frac{\omega}{\omega_0}\right)^2 + j \frac{\omega}{\omega_0} \frac{1}{Q}} \quad (6)$$

$$T_{BP}(j\omega) = H_{OBP} H_{BP}(j\omega) = \frac{H_{OBP} j \frac{\omega}{\omega_0}}{1 - \left(\frac{\omega}{\omega_0}\right)^2 + j \frac{\omega}{\omega_0} \frac{1}{Q}} \quad (7)$$

$$T_{HP}(j\omega) = H_{OHP} H_{HP}(j\omega) = \frac{H_{OHP} \left(j \frac{\omega}{\omega_0}\right)^2}{1 - \left(\frac{\omega}{\omega_0}\right)^2 + j \frac{\omega}{\omega_0} \frac{1}{Q}} \quad (8)$$

where $j = \sqrt{-1}$, ω is the angular frequency, ω_0 is the resonant angular frequency of the filter, Q is the quality factor of the filter, H_{OLP} is the LP filter gain when $\omega = 0$, QH_{OBP} is the BP filter gain when $\omega = \omega_0$, and H_{OHP} is the HP filter gain when $\omega \gg \omega_0$. If such quadratic filter sections are realized as two-capacitor RC-active networks in which the electrical parameters of the active devices are independent of frequency, it is possible to conclude that V_2/V_k must take the form

$$\frac{V_2}{V_k} = \xi_{HP}^k T_{HP}(j\omega) + \xi_{BP}^k T_{BP}(j\omega) + \xi_{LP}^k T_{LP}(j\omega) \quad (9)$$

where ξ_{HP}^k, ξ_{BP}^k, and ξ_{LP}^k are independent of frequency.

If all of the noise sources are uncorrelated, the total output noise voltage spectral density can be obtained by an rms addition of the contribution to the output noise voltage spectral density due to the various noise sources. Thus if e_{nk} is the noise voltage spectral density associated with the noise voltage source V_k, the total output noise voltage spectral density will be given by [7]

$$e_{no}^2 = \sum_k \left|\frac{V_2}{V_k}\right|^2 e_{nk}^2. \quad (10)$$

If use is made of the relationship [8]

$$|H_{HP}(j\omega)|^2 = \left(2 - \frac{1}{Q^2}\right) |H_{BP}(j\omega)|^2 - |H_{LP}(j\omega)|^2 + 1 \quad (11)$$

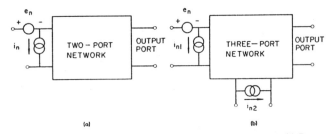

Fig. 2. (a) Representation of noise in a two-port network. (b) Representation of noise in a three-port network.

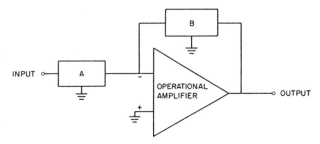

Fig. 3. Single-feedback operational amplifier circuit.

(9) and (10) can be combined to show that (10) will always take the form

$$e_{no}^2 = \phi_{BP}|H_{BP}(j\omega)|^2 + \phi_{LP}|H_{LP}(j\omega)|^2 + \phi_{AP}. \quad (12)$$

ϕ_{BP}, ϕ_{LP}, and ϕ_{AP} (the subscript AP refers to all-pass) are functions of the passive circuit elements, the frequency-independent electrical properties of the active devices, the noise sources associated with the passive elements, and the noise sources associated with the active elements.

An examination of (12) shows that $e_{no}^2 = \phi_{AP}$ for $\omega \to \infty$. The contribution to ϕ_{AP} by a noise source can therefore be calculated by short-circuiting the capacitors. Similarly, $e_{no}^2 = \phi_{LP} + \phi_{AP}$ when $\omega \to 0$. The sum of the contributions to ϕ_{LP} and ϕ_{AP} by any noise source can therefore be calculated by open-circuiting the capacitors. These observations provide a convenient method of checking the contributions to ϕ_{LP} and ϕ_{AP} when these are calculated with the aid of (9)–(11).

The total open-circuit output noise voltage in the frequency range f_l to f_h—$V_{no}(f_l,f_h)$—can be calculated by evaluating

$$V_{no}^2(f_l,f_h) = \int_{f_l}^{f_h} e_{no}^2 \, df. \quad (13)$$

Thus when the e_{nk} are independent of frequency, ϕ_{BP}, ϕ_{LP}, and ϕ_{AP} will be independent of frequency and only the two integrals

$$\int_{f_l}^{f_h} |H_{BP}(j\omega)|^2 \, df \quad (14)$$

and

$$\int_{f_l}^{f_h} |H_{LP}(j\omega)|^2 \, df \quad (15)$$

are required to calculate $V_{no}(f_l,f_h)$.

In practical two-capacitor RC-active networks, resistive noise and active device noise must be considered. The noise voltage spectral density e_{nR} that can be associated with a resistance R at the thermodynamic temperature T is

$$e_{nR}^2 = 4kTR \quad (16)$$

where k is Boltzmann's constant. The noise properties of two-port [9] and three-port [10] active networks can be represented by noise voltage and noise current sources as shown in Fig. 2. These noise voltage and noise current sources often have noise voltage and noise current spectral densities that can be described analytically in the low frequency range by

$$e_{na}^2 = e_{nav}^2 + \left(\frac{f_e}{f}\right) e_{nav}^2 \quad (17)$$

$$i_{na}^2 = i_{nav}^2 + \left(\frac{f_i}{f}\right) i_{nav}^2. \quad (18)$$

The noise voltage and noise current spectral densities e_{nav} and i_{nav} and the characteristic $1/f$ noise corner frequencies f_e and f_i would normally be determined by fitting (17) and (18) to experimental data. There would, in the general case, be some correlation between the noise voltage source and the noise current source or sources, and between the noise current sources. The present work, however, is concerned with the case when the noise sources are statistically independent so that (10) may be used.

Since a $1/f$ dependence of the power spectrum at low frequencies implies an unbounded total noise power, it should be understood that (17) and (18) must be restricted to the frequency above some arbitrary frequency f_d. This frequency f_d would be the dividing point between what should be considered to be noise and what should be considered to be dc drift in a particular application.

The output noise voltage in the frequency range from f_d to some upper frequency f_x would generally be of interest in the case of all three types of sections. The output noise in the frequency range f_1 to f_2 where f_1 and f_2 are the usual upper and lower 3-dB cutoff frequencies would be of interest in the BP case. If only white (independent of frequency) noise sources and $1/f$ noise sources are considered, the evaluation of (13) will involve the first nine integrals listed in the Appendix [11]–[13]. The approximations involved in extending the integrals to range from $f = 0$ to $f = \infty$ in (48), (50), (52), and (54) are relatively good ones. For example, if $(f_d/f_0) = 0.2$ and $(f_x/f_0) = 5$ in (48) and (52), the exact values of the integrals would be $1.37 f_0 Q$ rather than $1.57 f_0 Q$.

Single-feedback RC-active circuits of the type shown in Fig. 3 are frequently used to realize quadratic transfer functions having a complex pole pair. If the networks A and B have no more than two capacitors each and if the operational amplifier input is a virtual earth so that A and B are effectively separated as far as the admittance parameter calculations for each network are concerned, the technique for noise calculations that has been described above will be applicable. Thus many practical RC-active filter noise analyses can be performed in this way despite the seemingly very severe restriction to two-capacitor circuits.

TABLE I
COMPONENTS OF ϕ_{BP}, ϕ_{LP}, AND ϕ_{AP}—LP QUADRATIC FILTER SECTION

	ϕ_{BP}	ϕ_{LP}	ϕ_{AP}
R_1	0	$4kTR_1 H^2$	0
R_3	$4kTR_3 Q^2 \left(1 + H + \frac{R_4}{R_3}\right)^2$	$4kTR_3 (1+H)^2$	0
R_4	0	$4kTR_4$	0
R_6	$4kTR_6 \left[Q^2\left(1+H+\frac{R_4}{R_3}\right)^2 + 2\left(1+\frac{R_4}{R_3}\right)\right]$	$4kTR_6(2+H)H$	$4kTR_6$
e_{na}	$e_{na}^2 \left[Q^2\left(1+H+\frac{R_4}{R_3}\right)^2 + 2\left(1+\frac{R_4}{R_3}\right)\right]$	$e_{na}^2(2+H)H$	e_{na}^2
i_{na1}	$(i_{na1} R_3)^2 Q^2\left(1+H+\frac{R_4}{R_3}\right)^2$	$(i_{na1} R_3)^2 \left(1+H+\frac{R_4}{R_3}\right)^2$	0
i_{na2}	$(i_{na2} R_6)^2 \left[Q^2\left(1+H+\frac{R_4}{R_3}\right)^2 + 2\left(1+\frac{R_4}{R_3}\right)\right]$	$(i_{na2} R_6)^2 (2+H)H$	$(i_{na2} R_6)^2$

III. Examples Involving Operational Amplifier Realizations of Quadratic Sections

Multiple-feedback LP, BP, and HP quadratic sections realized with the aid of differential-input single-ended output operational amplifiers serve as a convenient vehicle for illustrating the procedure for noise analysis that has been outlined in Section II. The amplifiers are assumed to have infinite input impedance and zero output impedance. The voltage gain is assumed to be very high and independent of frequency. The contributions to ϕ_{BP}, ϕ_{LP}, and ϕ_{AP} due to e_{na}, i_{na1}, i_{na2}, and the noise sources associated with the resistors would be obtained as outlined in Section II. Since the results are complicated even for the very simplest circuits, a systematic scheme for tabulating these contributions has been adopted.

A. Multiple-Feedback LP Section

The LP single-amplifier multiple-feedback quadratic section shown in Fig. 4 has the voltage transfer function [14]

$$T_{LP}(j\omega) = \frac{-H}{1 - \left(\frac{\omega}{\omega_0}\right)^2 + j\left(\frac{\omega}{\omega_0}\right)\frac{1}{Q}} \quad (19)$$

where

$$H = \frac{R_4}{R_1} \quad (20)$$

$$\omega_0^2 = \frac{1}{C_2 C_5 R_3 R_4} \quad (21)$$

and

$$Q = \frac{\sqrt{\frac{C_2}{C_5}\frac{R_4}{R_3}}}{\left(1 + \frac{R_4}{R_1} + \frac{R_4}{R_3}\right)}. \quad (22)$$

The various contributions to ϕ_{BP}, ϕ_{LP}, and ϕ_{AP} are summarized in Table I. An examination of Table I shows that contributions due to e_{na} are independent of impedance level, con-

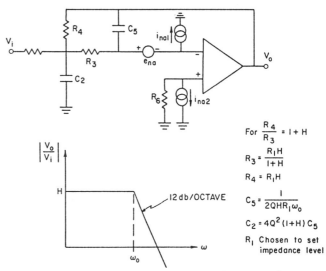

Fig. 4. LP multiple-feedback RC-active filter network.

For $\frac{R_4}{R_3} = 1 + H$

$R_3 = \frac{R_1 H}{1+H}$

$R_4 = R_1 H$

$C_5 = \frac{1}{2QHR_1\omega_0}$

$C_2 = 4Q^2(1+H)C_5$

R_1 Chosen to set impedance level

tributions due to R_1, R_3, R_4, and R_6 are proportional to the impedance level, and contributions due to i_{na1} and i_{na2} are proportional to the square of the impedance level. If a choice of impedance level is available then for a given Q, ω_0, H, and a given amplifier, the output noise level will be a minimum when the contribution to the output noise due to e_{na} is dominant. A specification of Q, ω_0, H, and the general impedance level of the circuit via R_1 still leaves one degree of freedom at the designer's discretion. A large capacitance ratio is undesirable and R_4/R_3 is sometimes chosen so that C_2/C_5 has the smallest possible value for a given Q and H. This occurs when $(R_4/R_3) = 1 + H$. Moreover, the resistor R_6 is usually chosen to be equal to $R_3 + R_1 R_4/(R_1 + R_4)$ to minimize dc drift due to amplifier input current variations. With the aid of these additional design constraints, Table I can be rewritten in terms of $Q, H, R_1, e_{na}, i_{na1}$, and i_{na2} only. The ratio of R_4/R_3 can, of course, be selected to minimize the output noise in some fashion, and Table I is thus a useful starting point for such minimization procedures.

It is useful to divide the output noise into an "inherent noise" component (that due to R_1, R_3, and R_4), and an "amplifier noise" component (that due to e_{na}, i_{na1}, i_{na2}, and

R_6) [8]. Since Johnson noise has a white spectrum, f_d can be taken to be zero in calculating the output noise due to R_1, R_3, and R_4. Thus for $f_x \gg f_0$, (13), (12), and Table I can be used to show that the inherent output noise of the circuit of Fig. 4 is given by

$$V_{noi}^2 = 4kTR_1 H \left[\left\{ Q^2 \left(1 + H + \frac{R_4}{R_3}\right)^2 + (1+H)^2 \right\} \frac{R_3}{R_4} + (1+H) \right] \frac{\pi}{2} f_0 Q. \quad (23)$$

If the design with $(R_4/R_3) = 1 + H$ is used, the inherent noise will be given by

$$V_{noi}^2 = 4kTR_1 H(1+H)(2Q^2 + 1)\pi f_0 Q. \quad (24)$$

B. Multiple-Feedback BP Section

The BP single-amplifier multiple-feedback quadratic section shown in Fig. 5 has the voltage transfer function [14]

$$T_{BP}(j\omega) = \frac{-H\left(j\frac{\omega}{\omega_0}\right)}{1 - \left(\frac{\omega}{\omega_0}\right)^2 + j\left(\frac{\omega}{\omega_0}\right)\frac{1}{Q}} \quad (25)$$

where

$$H = \sqrt{\frac{\frac{R_5}{R_1}\frac{C_3}{C_4}}{1 + \frac{R_1}{R_2}}} \quad (26)$$

$$\omega_0^2 = \frac{1}{C_4 R_5 \left(\frac{C_3 R_1 R_2}{R_1 + R_2}\right)} \quad (27)$$

and

$$Q = \frac{\sqrt{\frac{R_5 C_3}{R_1 C_4}\left(1 + \frac{R_1}{R_2}\right)}}{\left(1 + \frac{C_3}{C_4}\right)}. \quad (28)$$

The various contributions to ϕ_{BP}, ϕ_{LP}, and ϕ_{AP} are summarized in Table II, where

$$\gamma = 1 + Q^2\left(1 + \frac{C_3}{C_4}\right). \quad (29)$$

It will be noted that the contributions to ϕ_{BP}, ϕ_{LP}, and ϕ_{AP} due to e_{na}, i_{na1}, i_{na2}, and the resistors have the same dependence on the impedance level as in the LP filter section example.

The inherent output noise of the circuit of Fig. 5 is given by

$$V_{noi}^2 = 4kTR_1 H^2 \left[1 + \frac{R_1}{R_2}\right] \frac{\pi}{2} f_0 Q. \quad (30)$$

The value of Q is a maximum for a given value of $[R_5/(R_1 R_2/R_1 + R_2)]$ when the capacitors are selected so that $C_3/C_4 = 1$.

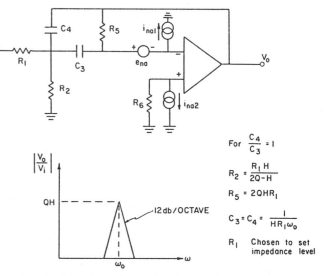

Fig. 5. BP multiple-feedback RC-active filter network.

TABLE II
COMPONENTS OF ϕ_{BP}, ϕ_{LP}, AND ϕ_{AP}–BP QUADRATIC FILTER SECTION

	ϕ_{BP}	ϕ_{LP}	ϕ_{AP}
R_1	$4kTR_1 H^2$	0	0
R_2	$4kTR_2 \left(\frac{R_1}{R_2}\right)^2 H^2$	0	0
R_5	$\frac{4kTR_5}{Q^2}$	$4kTR_5$	0
R_6	$4kTR_6 \frac{(\gamma^2 - 1)}{Q^2}$	0	$4kTR_6$
e_{na}	$e_{na}^2 \frac{(\gamma^2 - 1)}{Q^2}$	0	e_{na}^2
i_{na1}	$\frac{(i_{na1} R_5)}{Q^2}$	$(i_{na1} R_5)^2$	0
i_{na2}	$(i_{na2} R_6)^2 \frac{(\gamma^2 - 1)}{Q^2}$	0	$(i_{na2} R_6)^2$

Under these conditions

$$V_{noi}^2 = 4kTR_1 HQ^2 \pi f_0. \quad (31)$$

If the resistor R_6 is made equal to R_5 in order to minimize dc drift due to amplifier input current variations, the entries in Table II can be rewritten in terms of $Q, H, R_1, e_{na}, i_{na1}$, and i_{na2} only.

C. Multiple-Feedback HP Section

The HP single-amplifier multiple-feedback quadratic section shown in Fig. 6 has the voltage transfer function [14]

$$T_{HP}(j\omega) = \frac{-H\left[j\left(\frac{\omega}{\omega_0}\right)\right]^2}{1 - \left(\frac{\omega}{\omega_0}\right)^2 + j\left(\frac{\omega}{\omega_0}\right)\frac{1}{Q}} \quad (32)$$

where

$$H = \frac{C_1}{C_4} \quad (33)$$

TABLE III
COMPONENTS OF $\phi_{BP}, \phi_{LP},$ AND ϕ_{AP}—HP QUADRATIC FILTER SECTION

	ϕ_{BP}	ϕ_{LP}	ϕ_{AP}
R_2	$4kTR_2 Q^2 \left(1 + H + \dfrac{C_3}{C_4}\right)^2$	0	0
R_5	$\dfrac{4kTR_5}{Q^2}$	$4kTR_5$	0
R_6	$4kTR_6\left[\left(\dfrac{\gamma}{Q}\right)^2 + (1+H)^2\left(2 - \dfrac{1}{Q^2}\right) - 2(1+H)\right]$	$4kTR_6[1-(1+H)^2]$	$4kTR_6(1+H)^2$
e_{na}	$e_{na}^2\left[\left(\dfrac{\gamma}{Q}\right)^2 + (1+H)^2\left(2 - \dfrac{1}{Q^2}\right) - 2(1+H)\right]$	$e_{na}^2[1-(1+H)^2]$	$e_{na}^2(1+H)^2$
i_{na1}	$\dfrac{(i_{na1}R_5)^2}{Q^2}$	$(i_{na1}R_5)^2$	0
i_{na2}	$(i_{na2}R_6)^2\left[\left(\dfrac{\gamma}{Q}\right)^2 + (1+H)^2\left(2 - \dfrac{1}{Q^2}\right) - 2(1+H)\right]$	$(i_{na2}R_6)^2[1-(1+H)^2]$	$(i_{na2}R_6)^2(1+H)^2$

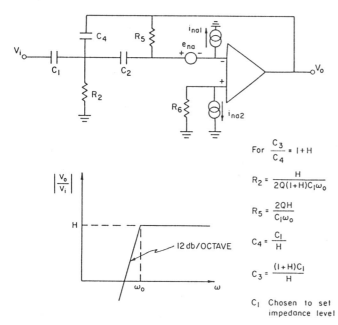

Fig. 6. HP multiple-feedback RC-active filter network.

$$\omega_0^2 = \frac{1}{C_4 R_2 C_3 R_5} \tag{34}$$

and

$$Q = \frac{\sqrt{\dfrac{C_3 R_5}{C_4 R_2}}}{\left(1 + \dfrac{C_1}{C_4} + \dfrac{C_3}{C_4}\right)}. \tag{35}$$

The circuit of Fig. 6 has three capacitors in it. However, because the negative input to the amplifier is a virtual earth, C_2 will be in parallel with C_1 during the y_{22} calculation and in parallel with C_4 during the y_{21} calculation. The y_{2k} calculations for $k > 1$ will all be done with the input and output short-circuited so that C_4 will be in parallel with C_1. Thus the three-capacitor circuit is effectively a two-capacitor circuit as far as noise calculation is concerned. The various contributions to $\phi_{BP}, \phi_{LP},$ and ϕ_{AP} are summarized in Table III, where

$$\gamma = 1 + Q^2\left(1 + \frac{C_1}{C_4} + \frac{C_3}{C_4}\right). \tag{36}$$

The contributions to $\phi_{BP}, \phi_{LP},$ and ϕ_{AP} due to e_{na}, i_{na1}, i_{na2}, and the resistors once again have the same dependence on the impedance level that has been noted for the case of the LP and BP filter section examples.

The inherent output noise of the circuit of Fig. 6 is given by

$$V_{noi}^2 = 4kTR_2\left(1 + \frac{C_1}{C_4} + \frac{C_3}{C_4}\right)^2\left[Q^2 + \frac{C_4}{C_3}(1+Q^2)\right]\frac{\pi}{2}f_0 Q. \tag{37}$$

The value of Q is a maximum for a given H and (R_5/R_2) when $(C_3/C_4) = 1 + (C_1/C_4)$. For this design

$$V_{noi}^2 = 8kTR_2[Q^2(1+H)^2 + (1+H)(1+Q^2)]\pi f_0 Q. \tag{38}$$

The resistor R_6 will normally be selected to be equal to R_5 in order to minimize dc drift due to amplifier input current variations. Thus the entries in Table III can be rewritten in terms of $Q, H, R_2, e_{na}, i_{na1},$ and i_{na2} only.

D. A Numerical Example—LP Case

Since Tables I, II, and III differ only in detail, the $(R_4/R_3) = (1 + H)$ LP design of Fig. 4 with $f_0 = 100$ Hz, $H = 10$, $Q = (1/\sqrt{2})$ (maximally flat case), $R_1 = 1125$ Ω, $R_3 = 1020$ Ω, $R_4 = 11.25$ kΩ, $R_6 = 2040$ Ω, $C_5 = 0.1$ μF, and $C_2 = 2.2$ μF is sufficient to illustrate a complete noise calculation for all three types of filter sections. Measurements of $e_{na}, i_{na1},$ and i_{na2} for the RCA CA3741CT, a bipolar input operational amplifier, yield

$$e_{na}^2 = e_{nav}^2\left(1 + \frac{f_e}{f_0} \cdot \frac{f_0}{f}\right)$$

$$\simeq 4.4 \times 10^{-16}\left(1 + 1.25 \frac{f_0}{f}\right) (V)^2/Hz \tag{39}$$

and

$$i_{na1}^2 = i_{na2}^2 = i_{nav}^2\left(1 + \frac{f_i}{f_0} \cdot \frac{f_0}{f}\right)$$

$$\simeq 1.5 \times 10^{-24} \frac{f_0}{f} (A)^2/Hz. \tag{40}$$

The entries in Table I can be evaluated to produce Table IV. If $f_d = 1$ Hz and $f_x = 10\,000$ Hz, (48)-(55) can be used to evaluate (13) to show that $V_{no}(f_x, f_d) = 8.32$-μV rms as compared to a measured value of 8.8-μV rms. The output noise due to the resistors R_1, R_3, R_4, and R_6 is 1.64-μV rms, that due to e_{na} is 8.10-μV rms, and that due to i_{na1} and i_{na2} is 0.927-μV rms. If the impedance level of the circuit is increased by a factor of ten by making all resistors larger by a factor of ten and all capacitors smaller by a factor of ten, the output noise due to the resistors R_1, R_3, R_4, and R_6 will be 5.19-μV rms, that due to e_{na} will remain at 8.10-μV rms, and that due to i_{na1} and i_{na2} will be 9.27-μV rms. The total output noise voltage would then be 13.4-μV rms.

IV. Tandem Combinations of Two Quadratic Sections

The foregoing analysis of single quadratic sections can be readily extended to include the case of filters produced by the tandem combination of two quadratic sections. The present discussion is limited to a consideration of a fourth-order maximally flat LP filter, but it should be evident that other types of fourth-order filters are also amenable to such analysis.

If an LP filter section labeled a is followed by an LP filter section labeled b to produce a fourth-order filter, the total noise voltage $V_{not}(f_d, f_x)$ at the output of section b in the frequency range f_d to f_x will be given by

$$V_{not}^2(f_d, f_x) = \int_{f_d}^{f_x} [\phi_{BPb} |H_{BPb}(j\omega)|^2$$
$$+ \phi_{LPb} |H_{LPb}(j\omega)|^2 + \phi_{APb}] \, df$$
$$+ \int_{f_d}^{f_x} [\phi_{BPa} |H_{BPa}(j\omega)|^2$$
$$+ \phi_{LPa} |H_{LPa}(j\omega)|^2 + \phi_{APa}]$$
$$\cdot [H_{0LPb}^2 |H_{LPb}(j\omega)|^2] \, df. \quad (41)$$

A maximally flat response [15] will be obtained if $\omega_{0a} = \omega_{0b} = \omega_0$, $Q_a = 0.541$ and $Q_b = 1.307$, or $Q_a = 1.307$ and $Q_b = 0.541$. In that case,

$$|H_{LPa}(j\omega) H_{LPb}(j\omega)|^2 = \frac{1}{1 + \left(\frac{f}{f_0}\right)^8}. \quad (42)$$

If white noise and $1/f$ noise sources are considered, the list of integrals represented by (48)-(56) in the Appendix will have to be augmented by (57)-(60) in order to evaluate V_{not}^2. Once tables like Table IV are prepared for section a and section b, the analysis of the fourth-order maximally flat LP filter is not significantly different from that of any other quadratic filter section.

As in the case of single sections, (41) may be used as a starting point for noise minimization procedures. For example, suppose the filter discussed in Section III-D is required to be a maximally flat fourth-order one with $f_0 = 100$ Hz, $H_t = H_a H_b = 10$, and $R_{1a} \simeq 1$ kΩ. For the $(R_4/R_3) = 1 + (R_4/R_1)$ design, amplifiers of the quality discussed in Section III-D, and the impedance level in question, the noise at the output due to e_{naa} and e_{nab} can be assumed to be dominant. It is then easy to show that, provided $e_{naa}^2 = e_{nab}^2 \simeq e_{na}^2$ of (39),

$$V_{not}^2 / e_{nav}^2 f_0 = [4Q_b^2(1+H_b)^2 + 2(2+H_b)] \rho_1$$
$$+ [2H_b(1+H_b)] \rho_2$$
$$+ [4Q_a^2(1+H_a)^2 + 2(2+H_a)] H_b^2 \rho_3$$
$$+ [(2+H_a) H_a H_b^2] \rho_4$$
$$+ \frac{f_x}{f_0} + \frac{f_e}{f_0} \ln \frac{f_x}{f_d} \quad (43)$$

where

$$\rho_1 = \frac{\pi}{2} Q_b + \frac{\frac{f_e}{f_0} Q_b}{\sqrt{1 - \frac{1}{4Q_b^2}}} \tan^{-1} 2Q_b \sqrt{1 - \frac{1}{4Q_b^2}} \quad (44)$$

$$\rho_2 = \frac{\pi}{2} Q_b + \frac{f_e}{f_0} \frac{Q_b \left[1 - \frac{1}{2Q_b^2}\right]}{\sqrt{1 - \frac{1}{4Q_b^2}}} \tan^{-1} 2Q_b \sqrt{1 - \frac{1}{4Q_b^2}} + \frac{f_e}{f_0} \ln \frac{f_0}{f_d} \quad (45)$$

$$\rho_3 = \frac{\frac{\pi}{8}}{\sin \frac{3\pi}{8}} + \frac{\pi \frac{f_e}{f_0}}{4\sqrt{2}} \quad (46)$$

and

$$\rho_4 = \frac{\frac{\pi}{8}}{\sin \frac{\pi}{8}} + \frac{\frac{f_e}{f_0}}{8} \ln \frac{f_x}{f_d} \frac{\left(1 + \frac{f_d}{f_0}\right)}{\left(1 + \frac{f_x}{f_0}\right)}. \quad (47)$$

Since ρ_1, ρ_2, ρ_3, and ρ_4 are all greater than zero, V_{not}^2 does not have a maximum for positive H_a and H_b. The value of V_{not}^2 is least when H_a is large and $Q_a = 0.541$. If $(f_d/f_0) = 0.01$, $(f_x/f_0) = 100$, and e_{naa}^2 and e_{nab}^2 are as given by (39), V_{not} will take on the values 6.65, 5.47, and 4.96-μV rms for $H_a = 5$, 10, and 20, respectively. If Q_a is taken to be 1.307, V_{not} will take on the values 8.45, 7.60, and 7.20 for $H_a = 5$, 10, and 20, respectively.

V. Summary

The quadratic filter section noise-analysis procedure that has been outlined in this work is very convenient for carrying out practical calculations and should prove to be useful in carrying out comparative studies of various realizations of a given transfer function. In addition, the noise performance of tandem combinations of quadratic filter sections can be handled in a straightforward manner. Indeed, an example has been given to show that the analysis of a fourth-order maximally flat LP filter can be carried out without recourse to numerical integration if the noise analysis of each quadratic section in the tandem combination is available.

TABLE IV
Components of ϕ_{BP}, ϕ_{LP}, and ϕ_{AP}—LP Quadratic Filter Section

	ϕ_{BP}	ϕ_{LP}	ϕ_{AP}
R_1	0	1.86×10^{-15}	0
R_3	4.09×10^{-15}	2.04×10^{-15}	0
R_4	0	1.86×10^{-16}	0
R_6	1.30×10^{-14}	4.06×10^{-15}	3.38×10^{-17}
e_{na}	$1.17 \times 10^{-13}\left(1 + 1.25\frac{f_0}{f}\right)$	$5.28 \times 10^{-14}\left(1 + 1.25\frac{f_0}{f}\right)$	$4.4 \times 10^{-16}\left(1 + 1.25\frac{f_0}{f}\right)$
i_{na1}	$3.78 \times 10^{-16}\frac{f_0}{f}$	$7.55 \times 10^{-16}\frac{f_0}{f}$	0
i_{na2}	$1.66 \times 10^{-15}\frac{f_0}{f}$	$7.51 \times 10^{-16}\frac{f_0}{f}$	$6.25 \times 10^{-18}\frac{f_0}{f}$

Note: All entries are in square volts per hertz.

Appendix

$$\int_{f_d}^{f_x} |H_{BP}(j\omega)|^2 \, df \simeq \int_0^{\infty} |H_{BP}(j\omega)|^2 \, df,$$

$$\text{for } \frac{f_d}{f_0} < 0.2, \frac{f_x}{f_0} > 5$$

$$= \frac{\pi}{2} f_0 Q \quad (48)$$

$$\int_{f_1}^{f_2} |H_{BP}(j\omega)|^2 \, df = \frac{\pi}{4} f_0 Q \quad (49)$$

$$\int_{f_d}^{f_x} \frac{f_0}{f} |H_{BP}(j\omega)|^2 \, df \simeq \int_0^{\infty} \frac{f_0}{f} |H_{BP}(j\omega)|^2 \, df,$$

$$\text{for } \frac{f_d}{f_0} < 0.2, \frac{f_x}{f_0} > 5$$

$$= \frac{Qf_0}{\sqrt{1 - \frac{1}{4Q^2}}} \tan^{-1}\left[2Q\sqrt{1 - \frac{1}{4Q^2}}\right],$$

$$Q > \frac{1}{2}$$

$$\simeq \frac{\pi}{2} f_0 Q, \quad Q > 4 \quad (50)$$

$$\int_{f_1}^{f_2} \frac{f_0}{f} |H_{BP}(j\omega)|^2 \, df = \frac{Qf_0}{\sqrt{1 - \frac{1}{4Q^2}}} \tan^{-1} \frac{\sqrt{1 - \frac{1}{4Q^2}}}{\sqrt{1 + \frac{1}{4Q^2}}},$$

$$Q > \frac{1}{2}$$

$$\int_{f_1}^{f_2} \frac{f_0}{f} H_{BP}(j\omega)|^2 \, df \simeq \frac{\pi}{4} f_0 Q, \quad Q > 2 \quad (51)$$

$$\int_{f_d}^{f_x} |H_{LP}(j\omega)|^2 \, df \simeq \int_0^{\infty} |H_{LP}(j\omega)|^2 \, df,$$

$$\text{for } \frac{f_d}{f_0} < 0.2, \frac{f_x}{f_0} > 5$$

$$\int_{f_d}^{f_x} |H_{LP}(j\omega)|^2 \, df = \frac{\pi}{2} f_0 Q \quad (52)$$

$$\int_{f_1}^{f_2} |H_{LP}(j\omega)|^2 \, df = \frac{\pi}{4} f_0 Q,$$

$$f_1 \text{ and } f_2 \text{ defined for } H_{BP}(j\omega) \quad (53)$$

$$\int_{f_d}^{f_x} \frac{f_0}{f} |H_{LP}(j\omega)|^2 \, df \simeq \frac{\left[1 - \frac{1}{2Q^2}\right] Qf_0}{\sqrt{1 - \frac{1}{4Q^2}}} \tan^{-1}\left[2Q\sqrt{1 - \frac{1}{4Q^2}}\right]$$

$$+ f_0 \ln\left(\frac{f_0}{f_d}\right), \quad Q > \frac{1}{2}, \frac{f_d}{f_0} < 0.2,$$

$$\frac{f_x}{f_0} > 5$$

$$\simeq \frac{\pi}{2} f_0 Q + f_0 \ln\left(\frac{f_0}{f_d}\right), \quad Q > 4,$$

$$\frac{f_d}{f_0} < 0.2, \frac{f_x}{f_0} > 5 \quad (54)$$

$$\int_{f_1}^{f_2} \frac{f_0}{f} |H_{LP}(j\omega)|^2 \, df = \frac{\left[1 - \frac{1}{2Q^2}\right]}{\sqrt{1 - \frac{1}{4Q^2}}} Qf_0 \tan^{-1} \frac{\sqrt{1 - \frac{1}{4Q^2}}}{\sqrt{1 + \frac{1}{4Q^2}}}$$

$$+ \frac{f_0}{2} \ln\left(\frac{f_2}{f_1}\right), \quad Q > \frac{1}{2},$$

$$f_1 \text{ and } f_2 \text{ defined for } H_{BP}(j\omega)\omega$$

$$\simeq \frac{\pi}{4} f_0 Q + \frac{f_0}{2} \ln \frac{\sqrt{4Q^2 + 1} + 1}{\sqrt{4Q^2 + 1} - 1},$$

$$Q > 2 \quad (55)$$

$$\int_{f_d}^{f_x} \frac{f_0}{f} \, df = f_0 \ln \frac{f_x}{f_d} \quad (56)$$

$$\int_{f_d}^{f_x} \frac{\left(\frac{f}{f_0}\right)^2 df}{1+\left(\frac{f}{f_0}\right)^8} \simeq \int_0^\infty \frac{\left(\frac{f}{f_0}\right)^2 df}{1+\left(\frac{f}{f_0}\right)^8},$$

$$\text{for } \frac{f_d}{f_0} < 0.2 \text{ and } \frac{f_x}{f_0} > 5$$

$$= \frac{\frac{\pi}{8} f_0}{\sin\left(\frac{3\pi}{8}\right)} \quad (57)$$

$$\int_{f_d}^{f_x} \frac{\left(\frac{f}{f_0}\right) df}{1+\left(\frac{f}{f_0}\right)^8} \simeq \int_0^\infty \frac{\left(\frac{f}{f_0}\right) df}{1+\left(\frac{f}{f_0}\right)^8},$$

$$\text{for } \frac{f_d}{f_0} < 0.2 \text{ and } \frac{f_x}{f_0} > 5$$

$$= \frac{\pi f_0}{4\sqrt{2}} \quad (58)$$

$$\int_{f_d}^{f_x} \frac{df}{1+\left(\frac{f}{f_0}\right)^8} \simeq \int_0^\infty \frac{df}{1+\left(\frac{f}{f_0}\right)^8},$$

$$\text{for } \frac{f_d}{f_0} < 0.2 \text{ and } \frac{f_x}{f_0} > 5$$

$$= \frac{\frac{\pi}{8} f_0}{\sin\frac{\pi}{8}} \quad (59)$$

$$\int_{f_d}^{f_x} \frac{\left(\frac{f_0}{f}\right) df}{1+\left(\frac{f}{f_0}\right)^8} = \frac{f_0}{8} \ln \frac{f_x}{f_d} \frac{\left(1+\frac{f_d}{f_0}\right)}{\left(1+\frac{f_x}{f_0}\right)}$$

$$\simeq \frac{f_0}{8} \ln \frac{f_0}{f_d}, \quad \frac{f_d}{f_0} < 0.2 \text{ and } \frac{f_x}{f_0} > 5. \quad (60)$$

References

[1] W. R. Bennett, *Electrical Noise*. New York: McGraw-Hill, 1960.
[2] B. A. Shenoi, "Practical realization of a gyrator circuit and *RC*-gyrator filters," *IEEE Trans. Circuit Theory*, vol. CT-12, pp. 374–380, Sept. 1965.
[3] L. T. Bruton, "Nonideal performance of two-amplifier positive-impedance converters," *IEEE Trans. Circuit Theory*, vol. CT-17, pp. 541–549, Nov. 1970.
[4] R. P. Sallen and E. L. Key, "A practical method of designing *RC* active filters," *IRE Trans. Circuit Theory*, vol. CT-2, pp. 74–85, Mar. 1955.
[5] S. K. Mitra, *Analysis and Synthesis of Linear Active Networks*. New York: Wiley, pp. 474–475.
[6] R. E. Bach, "Selecting *RC* values for active filters," *Electronics*, vol. 33, pp. 82–85, May 1960.
[7] H. C. Montgomery, "Transistor noise in circuit applications," *Proc. IRE*, vol. 40, pp. 1461–1471, Nov. 1952.
[8] L. T. Bruton, F. N. Trofimenkoff, and D. H. Treleaven, "Noise performance of low-sensitivity active filters," *IEEE J. Solid-State Circuits*, vol. SC-8, pp. 85–91, Feb. 1973.
[9] W. R. Huber, "Two-port equivalent noise generators," *Proc. IEEE* (Lett.), vol. 58, pp. 807–809, May 1970.
[10] D. H. Sheingold and L. Smith, "Operational-amplifier-circuit noise characteristics," *Electron. Inst. Digest*, pp. 50–57, Sept. 1969.
[11] F. N. Trofimenkoff, "Noise margins of band-pass filters," *IEEE Trans. Circuit Theory*, to be published.
[12] E. F. Vandivere, "Noise output of a multipole filter relative to that of the ideal square-response filter," *Proc. IEEE* (Corresp.), vol. 51, p. 1771, Dec. 1963.
[13] I. S. Gradshteyn and I. M. Ryzhik, *Table of Integrals, Series and Products*. New York: Academic, 1965.
[14] *Handbook of Operational Amplifier Active RC Networks*, Burr-Brown Research Corp., Tucson, Ariz., 1966.
[15] L. Weinberg, *Network Analysis and Synthesis*, New York: McGraw-Hill, 1962.

Noise Performance Limitations of Single Amplifier RC Active Filters

JAMES W. HASLETT, MEMBER, IEEE

Abstract—The noise performance characteristics of single amplifier RC active filters are presented for the cases where the circuits are constructed using the new single-ended Norton quad amplifiers. The power spectral densities and total rms values of the output noise are compared with the same quantities for filters constructed using conventional integrated operational voltage amplifiers. The conditions for minimum noise from the Norton amplifiers are derived, and these minimum levels are compared with those obtained using 741-type amplifiers. When the filters are constructed using infinite gain embedded amplifiers, for certain ranges of Q-factor and passband gain the Norton amplifiers can be biased to give comparable noise performance to the 741 amplifiers, and it is shown that the bias networks required differ significantly from those chosen in most specification sheets for these devices. For positive-gain structures, the 741 amplifiers are shown to be superior in most applications.

I. INTRODUCTION

THE INTRODUCTION of the new quad Norton single-ended integrated amplifiers has opened the way for some cheap compact signal processing systems. In particular, all of the single amplifier active filter networks can be readily constructed using these amplifiers. The noise performance of the networks is therefore of interest, and a comparison of noise levels from these filters with noise levels from identical filters constructed with differential voltage amplifiers is of some importance.

The noise performance of the National LM3900 quad series of amplifiers has recently been modeled by the author [1], and the performance of the differential voltage type of amplifier is well known [2]. The noise performance of single-amplifier RC quadratic filters has been analyzed in detail by Trofimenkoff, Treleaven, and Bruton [3], and the positive gain structure noise performance has been analyzed by Trofimenkoff, Smallwood, and Bruton [4]. Their results can be used to obtain the noise characteristics of the same filters constructed using Norton amplifiers, so that direct comparisons of circuit performance can easily be made.

II. RC ACTIVE QUADRATIC SECTIONS

The noise performance characteristics of these filters have been analyzed in detail by Trofimenkoff *et al.* [3] to account for a noise voltage (e_{nv}) and current generator (i_{nv1}) at the inverting input, and a noise current generator (i_{nv2}) at the noninverting input of the differential amplifier. The analysis is completely general in nature and is presented in such a

Manuscript received July 20, 1974; revised March 25, 1975. This work was supported by the National Research Council of Canada.
The author is with the Department of Electrical Engineering, the University of Calgary, Calgary, Alberta, Canada.

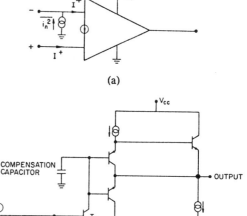

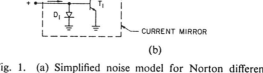

Fig. 1. (a) Simplified noise model for Norton differential current amplifier. (b) Schematic diagram of Norton amplifier.

way that the contributions to the output noise from each noise generator are given separately. When normal differential voltage amplifiers are used, the magnitudes of the noise generators are independent of the impedance and bias levels chosen for the network, and it can be shown that the noise characteristics are usually dominated by e_{nv} if the impedance level of the circuit is chosen properly [3]. When Norton amplifiers are used, on the other hand, the noise characteristics are dominated by a noise current generator i_{ni} at the inverting input, whose magnitude does depend on the impedance level chosen for the circuit. The magnitude of the noise generator has been calculated by the author [1], and the equivalent noise model is shown in Fig. 1.

The output power spectral densities of the three types of filters (low-pass, bandpass, and high-pass) can be easily calculated when differential voltage and current amplifiers are used simply by combining the appropriate terms from the results presented by Trofimenkoff *et al.* [3]. If only a noise voltage generator e_{nv} is active in producing noise, it can be shown that in all cases the power spectral density

Reprinted from *IEEE Trans. Circuits Syst.*, vol. CAS-22, no. 9, pp. 743–747, Sept. 1975.

TABLE I
A SUMMARY OF DESIGN EQUATIONS AND COEFFICIENTS ASSOCIATED WITH OUTPUT NOISE EQUATIONS FOR THREE SINGLE AMPLIFIER FILTER STRUCTURES

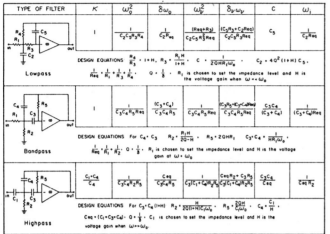

curves take on the form

$$e_{nov} = K \frac{e_{nv}(s^2 + \delta_v \omega_v s + \omega_v^2)}{(s^2 + \delta \omega_0 s + \omega_0^2)}. \quad (1)$$

The values of K, δ_v, and ω_v are different for each of the three filter types, and all parameters in (1) are presented in Table I.

Similar results are obtained when only a noise current generator is active. In these cases, the power spectral density curves are always of the form

$$e_{noi} = \frac{\frac{i_{ni}}{C}(s + \omega_i)}{(s^2 + \delta \omega_0 s + \omega_0^2)} \quad (2)$$

and the parameters C and ω_i are also given in Table I. (Note that δ and ω_0 are similar in (1) and (2) and are determined by the electrical characteristics of the filter under consideration.)

The power spectral densities of the output noise from each type of amplifier are then directly obtainable from (1) and (2). A comparison of power spectral density relationships is most useful in the passband of the filter being constructed. Direct comparison can be obtained by assuming that uncorrelated noise voltage and current generators at the inverting node of the differential voltage amplifier contribute to the output noise as mean-squared sums of (1) and (2), while only a noise current generator at the inverting node of the differential current amplifier contributes to the output noise in the form given by (2). (The noise contribution from the current generator at the noninverting node of the differential voltage amplifier is neglected, since it can always be eliminated if desired.) A comparison of the output power spectral densities for the two types of amplifiers yields the general result that

$$\left.\frac{\overline{e_{nov}^2}}{\overline{e_{noi}^2}}\right| = \frac{\overline{i_{nv}^2}}{\overline{i_{ni}^2}} + \frac{K^2 \overline{e_{nv}^2}\left\{\left(\frac{\omega_v^2}{\omega^2} - 1\right)^2 + \frac{\delta_v^2 \omega_v^2}{\omega^2}\right\}}{\frac{\overline{i_{ni}^2}}{\omega^2 C^2}\left[1 + \left(\frac{\omega_i}{\omega}\right)^2\right]} \quad (3)$$

and the ratio can be examined for each type of filter using the results given in Table I.

a) Low-Pass Case: The noise levels in the passband ($\omega \ll \omega_0$) are compared by taking $\lim_{\omega \to 0}$ (3) and substituting for ω_v, ω_i, K, and C to give

$$\left.\frac{\overline{e_{nov}^2}}{\overline{e_{noi}^2}}\right|_{\omega=0} = \frac{\overline{i_{nv}^2}}{\overline{i_{ni}^2}} + \frac{\overline{e_{nv}^2}}{\overline{i_{ni}^2} R_3^2}\left(1 + \frac{R_{eq}}{R_3}\right)^2. \quad (4)$$

b) Bandpass Case: A similar result to (4) is obtained for the bandpass filter by setting $\omega = \omega_0$ and $C_3 = C_4$ (normal design values) in (3) to give

$$\left.\frac{\overline{e_{nov}^2}}{\overline{e_{noi}^2}}\right|_{\omega=\omega_0} = \frac{\overline{i_{nv}^2}}{\overline{i_{ni}^2}} + \frac{\overline{e_{nv}^2}}{\overline{i_{ni}^2} R_5 R_{eq}} \frac{\left(1 + \frac{2R_{eq}}{R_5}\right)^2}{\left(1 + \frac{4R_{eq}}{R_5}\right)}. \quad (5)$$

c) High-Pass Case: For the case where $\omega > \omega_0$, (3) becomes

$$\left.\frac{\overline{e_{nov}^2}}{\overline{e_{noi}^2}}\right|_{\omega \gg \omega_0} = \frac{\overline{i_{nv}^2}}{\overline{i_{ni}^2}} + \frac{K^2 \overline{e_{nv}^2}(\omega C)^2}{\overline{i_{ni}^2}}. \quad (6)$$

In general, $\overline{e_{nv}^2}$ is independent of the bias conditions and impedance levels chosen for the filter, while $\overline{i_{ni}^2}$ is dependent on these parameters according to the relationship [1]

$$\overline{i_{ni}^2} = 4qI^+ \cdot \Delta f \left(1 + \frac{r_{bb}' qI^+}{kT}\right). \quad (7)$$

The current I^+ is determined by the impedance level chosen for the filter and the offset voltage required at the output. Once the impedance level of the filter has been defined, the resistance of the dc feedback path is known. If it is desired that the offset voltage at the output be centered midway between supply and ground, then the bias resistor associated with the amplifier must be twice as large as the feedback resistance and I^+ is determined accordingly [1].

For the low-pass and bandpass filters, the ratios R_{eq}/R_3 and R_{eq}/R_5 will be determined by the transfer function but will be independent of the impedance levels associated with the network. Thus the ratio $\overline{e_{nov}^2}/\overline{e_{noi}^2}$ can be varied for the low-pass case by the term $\overline{i_{ni}^2} R_3^2$, for the bandpass case by $\overline{i_{ni}^2} R_5 R_{eq}$, and for the high-pass case by $\overline{i_{ni}^2}/(\omega^2 C^2)$. From (7) it is apparent that there are two possible regions of operation: one in which $\overline{i_{ni}^2}$ is proportional to I^+, and another, at higher current levels where $\overline{i_{ni}^2}$ is proportional to $(I^+)^2$. In the low-pass case, if the input is ac coupled

$$I^+ = \frac{V_{cc}/2}{(R_3 + R_4)}. \quad (8)$$

Since R_3 and R_4 must always remain in a fixed ratio if the transfer function is to remain unchanged (i.e., $R_4 = R_3(1 + H)$), then

$$R_3 = \frac{V_{cc}}{2I^+(2 + H)} \quad (9)$$

or R_3^2 is proportional to $(I^+)^{-2}$. An inspection of the product $\overline{i_{ni}^2}R_3^2$ then reveals that at large values of R_3 (corresponding to low current levels) where $\overline{i_{ni}^2}$ is proportional to I^+, that

$$\overline{i_{ni}^2}R_3^2 \alpha \frac{1}{I^+} \alpha R_3. \tag{10}$$

Equation (10) indicates that increasing the impedance level of the circuit to minimize $\overline{i_{ni}^2}$ actually has a detrimental effect on the noise performance of the filter if a Norton amplifier is used. The noise is in fact minimized when large values of I^+ (and consequently a low impedance level for the circuit) are used, to bring $\overline{i_{ni}^2}$ into a range where it is proportional to $(I^+)^2$. The term $\overline{i_{ni}^2}R_3^2$ then becomes independent of I^+ and is minimized, and it is easy to show that this occurs when

$r_{bb}'qI^+/kT \gg 1$, which corresponds to $R_3 \ll \dfrac{r_{bb}'qV_{cc}}{2kT(2+H)}$. (11)

Typically $r_{bb}' = 300\ \Omega$ [1], $kT/q = 0.026$ V and if $V_{cc} = 20$ V, $H = 1$, then (11) corresponds to $R_3 \ll 40$ kΩ. This result is significant from the point of view of circuit design, since the present design values shown in the applications data indicate R_3 to be in the range of hundreds of kilohms. It should also be noted that since $\overline{i_{ni}^2}$ is large, the term $\overline{i_{nv}^2}/\overline{i_{ni}^2}$ is small and can be neglected when doing the comparison in (4).

It is of interest to determine an absolute value for the ratio (4). For the conventional design, it can be shown that $R_3 = 2R_{eq}$ independently of the value of H chosen. Then (4) becomes

$$\left.\frac{\overline{e_{nov}^2}}{\overline{e_{noi}^2}}\right|_{\omega=0} = \frac{9\overline{e_{nv}^2}/\Delta f}{4R_3^2\overline{i_{ni}^2}/\Delta f} = \frac{9\overline{e_{nv}^2}/\Delta f}{4q^2r_{bb}'V_{cc}^2/[kT(2+H)^2]}. \tag{12}$$

Typically, $V_{cc} = 20$ V, $H = 1$, $(\overline{e_{nv}^2}/\Delta f)^{1/2} = 30$ nV/$\sqrt{\text{Hz}}$ and (12) has a magnitude of about $1/4 \times 10^{-1}$. This means that the noise power at the output of a *unity gain* low-pass filter constructed with the LM3900 will be at least an order of magnitude larger than it will be for the same filter built using a 741 amplifier, when the noise outputs are minimized in both cases.

It is most important to note that decreasing V_{cc} or increasing H will have the effect of swinging (12) in favor of a Norton amplifier. Of course, when these parameters are varied the impedance level of the circuit must be altered so that (11) is still satisfied. If $V_{cc} = 4$ V (the minimum value), then for $H > 1.83$ the filter will be less noisy when constructed with an LM3900 amplifier. (If one could be certain that the 741 would have an equivalent noise voltage smaller than 30 nV/$\sqrt{\text{Hz}}$, then H would have to be larger before the tradeoff would occur. Such large values of H are usually not used in practice, and the 741 will make a less noisy low-pass filter for most applications.)

Similar arguments to those presented above can be put forth for the case of bandpass and high-pass filters. For the bandpass case, using the design equations given in Table I, it is easy to show that $R_{eq}/R_5 = (4Q^2)^{-1}$ so that (5) becomes

$$\left.\frac{\overline{e_{nov}^2}}{\overline{e_{noi}^2}}\right|_{\omega=\omega_0} \doteq \frac{\overline{e_{nv}^2}\left(1+\dfrac{1}{2Q^2}\right)^2 \cdot 4Q^2}{\overline{i_{ni}^2}R_5^2\left(1+\dfrac{1}{Q^2}\right)}. \tag{13}$$

In this case R_5 is the only dc feedback resistor, and since it determines the output offset level, in general [1]

$$R_5 \doteq \frac{V_{cc}}{2I^+}. \tag{14}$$

The ratio (13) is then maximized when $R_5 \ll r_{bb}'qV_{cc}/2kT$ which is similar to (11) except for the term $(2+H)$.

Under these conditions, (13) becomes

$$\left.\frac{\overline{e_{nov}^2}}{\overline{e_{noi}^2}}\right|_{\omega=\omega_0} = \frac{[4Q^2\overline{e_{nv}^2}/\Delta f]\left(1+\dfrac{1}{2Q^2}\right)^2 kT}{q^2r_{bb}'V_{cc}^2\left(1+\dfrac{1}{Q^2}\right)}. \tag{15}$$

Normally $Q^2 \gg 1$ so that (15) differs from the low-pass case by a factor $16Q_{BP}^2/[9(2+H_{LP})^2]$. From (15), if $V_{cc} = 4$ V, $Q > 8.5$ is the point at which better noise performance is obtained at $\omega = \omega_0$ with the Norton amplifier, independently of the value of H chosen for the network. This value of Q represents an upper limit on what can be obtained with this type of amplifier, so that the Norton amplifier displays poorer noise performance in general for any bandpass filter of this type, at the center frequency.

For the high-pass filter structure, (6) indicates that the ratio is frequency dependent, increasing as the frequency increases. Again, R_5 determines the offset voltage, and (14) is valid. If it is noted from the design procedure that $K = 1 + H$, and $C = Q/R_5\omega_0$, then (6) becomes

$$\left.\frac{\overline{e_{nov}^2}}{\overline{e_{noi}^2}}\right|_{\omega \gg \omega_0} = \frac{(1+H)^2\overline{e_{nv}^2}/\Delta f\left(\dfrac{\omega}{\omega_0}\right)^2 Q^2}{1.85 \times 10^{-15}V_{cc}^2}. \tag{16}$$

If $V_{cc} = 4$ V, $H = 1$, and $\omega \geq 5\omega_0$, then for $Q > 0.6$ the noise will be lower in the passband when an LM3900 amplifier is used. For larger values of H, it will be better to use a Norton amplifier in nearly all practical cases of interest.

Comparison of RMS Noise Levels

It is also of interest to compare total rms noise at the output when the two different types of amplifiers are used. The appropriate integrals have been published in the literature [3] and can easily be combined to produce expressions for the total mean-square output noise. When differential voltage amplifiers are used, the integrals theoretically diverge as the limits of integration are expanded; however physically meaningful results can be obtained for $f_h < Af_0$, where f_h is the upper frequency limit of interest and $\infty > A > 1$. (The noise current generator associated with the differential voltage amplifier will be ignored in the following discussion.) When Norton amplifiers are used, the integrals always converge due to the decreasing nature of the power spectral density curves with increasing frequency.

For the low-pass case, the ratio of mean-square output noise levels can be written as

$$\frac{\overline{V_{onv}^2}}{\overline{V_{oni}^2}} = \frac{\overline{e_{nv}^2}[4Q^3 \cdot (1+H)^2 + (2+H)^2 Q + K]}{(\overline{i_{ni}^2} \cdot R_3^2) \cdot 2(1+H) \cdot Q \cdot (1+Q^2)} \quad (17)$$

where $K = (\pi/2)(f_h/f_0)$. This is identical in form with (4), and the comparison of $\overline{e_{nv}^2}$ with $\overline{i_{ni}^2} R_3^2$ has already been carried out. It is apparent from (17) that the function is sensitive to Q for values of Q smaller than 0.5, and relatively independent of Q for larger values.

The actual value of the ratio depends, of course, on the values of K and R_3 chosen.

For the bandpass filter, the ratio can be written as

$$\frac{\overline{V_{onv}^2}}{\overline{V_{oni}^2}} = \frac{\overline{e_{nv}^2}[4(Q^3 + Q) + K]}{\overline{i_{ni}^2} \cdot R_1^2 4H^2 Q \cdot (Q^2 + 1)}. \quad (18)$$

This is identical in form to (13) if R_5 is replaced by $2QHR_1$. At larger values of Q it should be noted that (18) reduces to (13) as expected, since the mean-square noise approaches the spectral density in that narrow bandwidth.

For the high-pass filter, the ratio is written as

$$\frac{\overline{V_{onv}^2}}{\overline{V_{oni}^2}} \frac{\overline{e_{nv}^2} C_1^2 \omega_0^2}{\overline{i_{ni}^2}} = \frac{[(1+Q^2) + (1+H)2Q^2 + (1+H)^2 \cdot (4Q^4 + Q^2 + KQ - 1)]}{4Q^2 H^2 \cdot (Q^2 + 1)}. \quad (19)$$

In this case, it should be noted that the function will go negative for low values of Q unless K is kept large. This arises because at low values of Q the integrals associated with the magnitudes of the low-pass and bandpass functions are invalid unless $f_h \gg f_0$ which corresponds to $K \gg 1$. (This is true to some extent with the other filters as well, and one should take $0.2 < Q < 5$ as the normal region of operation.) Since (19) is an increasing function with increasing K, the ratio is only meaningful with K restricted to be greater than 5, but less than infinity. The conditions under which one amplifier is superior to another depend on the values of C_1 and ω_0 chosen for the network, and each filter constructed must be examined separately in this case.

It must be noted that the above results have been obtained by arbitrarily choosing an upper frequency limit to the integration of (1) when calculating total mean-square noise. Theoretically, the integrals diverge and the ratios (17)–(19) are meaningless unless bandlimited noise sources are assumed.

Single Amplifier Positive-Gain Structures

The noise performance of positive-gain-type structures has been analyzed by Trofimenkoff *et al.* [4] for the case where only an amplifier noise voltage generator $\overline{e_{na}^2}$ is active. The embedded positive gain amplifier associated with these filters is usually constructed with a differential

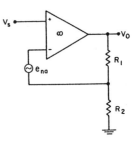

Fig. 2. Follower with gain constructed using differential voltage amplifier.

amplifier as shown in Fig. 2. The noise at the output is determined by the elements of the feedback network, and the equivalent noise voltage at the input is given by

$$\overline{e_{nv}^2} = \overline{e_{na}^2} + \frac{4kTR_1 \, \Delta f}{(1 + R_1/R_2)}. \quad (20)$$

Usually the impedance levels are kept low enough so that

$$\overline{e_{nv}^2} \simeq \overline{e_{na}^2}.$$

(When $K = 1$, $R_1 = 0$ and this is true regardless of the impedance level chosen for the filter.)

A similar "follower with gain" can be constructed using a Norton amplifier. A comparison of the noise performance of the two circuits is then achieved simply by calculating an equivalent noise voltage for the Norton follower, and comparing it with (20). The Norton circuit is constructed as shown in Fig. 3. The bias current I^+ is given by

$$I^+ = [V_{cc} - V_{D(on)}]/R_5 \quad (21)$$

where $V_{D(on)}$ is the dc voltage appearing across the input diode when I^+ flows into the input circuit. Since r_d, the small signal impedance of the diode, is given by kT/qI^+, then if $V_{cc} \gg V_{D(on)}$,

$$r_d \simeq \frac{kT}{q} \frac{R_5}{V_{cc}}. \quad (22)$$

The supply voltage V_{cc} is much greater than kT/q (26 mV at 300 K) under all conditions, so that $r_d \ll R_5$. Now R_3, R_4, and R_5 are normally of the same order of magnitude, and this gives

$$\frac{v_0}{v_s} \simeq \frac{R_4}{R_3} \quad (23)$$

under any bias conditions. Straightforward analysis of the equivalent noise voltage associated with the network gives

$$\overline{e_{ni}^2} = R_3^2 \left[\overline{i_{ni}^2} + 4kT \, \Delta f \left(\frac{1}{R_3} + \frac{1}{R_4} + \frac{1}{R_5} \right) \right]. \quad (24)$$

For a gain of K', the relationships $R_1 = K'R_2$, $R_5 = 2R_4$, $R_4 = K'R_3$ will hold, and a comparison of the noise levels

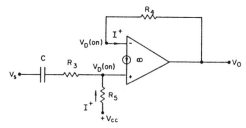

Fig. 3. Follower with gain constructed using differential current amplifier.

associated with both amplifiers gives

$$\frac{\overline{e_{nv}^2}}{\overline{e_{ni}^2}} = \frac{\overline{e_{na}^2} + \dfrac{K'}{1+K'} 4kTR_2 \, \Delta f}{\overline{i_{ni}^2} R_3^2 + \left(1 + \dfrac{1}{K'} + \dfrac{1}{2K'}\right) 4kTR_3 \, \Delta f}. \quad (25)$$

With the previous networks, we have attempted to minimize $\overline{i_{ni}^2} R_3^2$ by lowering the impedance level of the network. In this case, there are restrictions on the value of R_3 that can be chosen, since R_3 essentially defines the input impedance to the gain block K'. Usually it is desirable to maintain the impedance at a level in the megohm range. This puts $\overline{i_{ni}^2} R_3^2$ at a large level [1] and in (25) if the terms including K' are neglected (a justifiable assumption) then the ratio involves a direct comparison of $\overline{e_{na}^2}$ and $\overline{i_{ni}^2} R_3^2$. If $R_3 = 1$ MΩ and $K' = 1$, $V_{cc} = 4$ V minimizes (7) and $(\overline{i_{ni}^2} R_3^2)^{1/2} = 1$ μV/$\sqrt{\text{Hz}}$. This means that a filter constructed using the Norton amplifier will have a noise power output nearly two orders of magnitude higher than one constructed with something like a 741 differential voltage amplifier. The limitation is a serious one and is aggravated by the fact that the Norton follower should be ac coupled for ease of design. (If one is prepared to juggle the elements this can be overcome.) One other limitation occurs with the Norton amplifier which could cause problems in complex networks. The signal modulates the bias current flowing into the noninverting terminal of the amplifier when it is operated as a follower [1]. At larger signal levels, then, the noise sources are modulated by the signal in such a way that the overall noise level at the output becomes dependent on signal level. (The circuit is essentially a multiplier, multiplying the signal with white noise.) This could cause problems if noise levels at frequencies other than the signal frequency are of interest.

III. Conclusions

The noise performance characteristics of single amplifier RC active filters have been examined for the cases where a single-ended "Norton" differential current amplifier and a differential voltage amplifier are employed in the circuit construction. The noise levels associated with each are compared, and the conditions under which one amplifier is superior to another are examined. It is found that for filters which utilize an infinite gain embedded amplifier, the noise performance of both types is comparable and can be minimized by decreasing the impedance levels associated with the networks. This implies high bias current levels for the Norton amplifiers, and shows that the data sheet element values chosen for example filter designs using the Norton amplifiers lead to very noisy networks. For positive-gain structures, it is found that the Norton amplifier is a poor choice compared to a differential voltage amplifier, if impedance levels are kept high to maintain good electrical characteristics when a Norton amplifier is used.

References

[1] J. W. Haslett, "Noise performance of the new Norton op amps," *IEEE Trans. Electron Devices*, vol. ED-21, pp. 571–577, Sept. 1974.
[2] D. H. Sheingold and L. Smith, "Operational amplifier circuit noise characteristics," *Electron. Inst. Dig.*, pp. 50–57, Sept. 1969.
[3] F. N. Trofimenkoff, D. H. Treleaven, and L. T. Bruton, "Noise performance of RC-active quadratic filter sections," *IEEE Trans. Circuit Theory*, vol. CT-20, pp. 524–532, Sept. 1973.
[4] F. N. Trofimenkoff, R. E. Smallwood, and L. T. Bruton, "Noise in positive gain Sallen–Key RC active filters," presented at the 1974 European Conference on Circuit Theory, London, England, July 1974.

Part V: Noise In Switched-Capacitor Filters

Noise Sources and Calculation Techniques for Switched Capacitor Filters

JONATHAN H. FISCHER

Abstract—The noise response of switched capacitor networks (SCN's) is reviewed with emphasis on simplifying approximations suitable for SPICE noise simulation. The techniques developed cover all op-amp noise sources, as well as capacitor switching noise. The close agreement between predicted and measured noise responses for several monolithic SCN's bears out the validity of these simulation techniques.

I. Introduction

THE noise response derivations for switched capacitor networks (SCN's) developed in this paper differ from earlier work [1]-[3] by emphasizing approximations that facilitate the use of general-purpose programs, such as SPICE, for accurate SCN noise analysis. The derivations will cover ideal sampling effects, the development of a suitable SCN integrator noise model, and computer simulation techniques.

Applying the noise model to practical SCN's shows that the relative contributions of op-amp $1/f$, foldover flat-band, and capacitor switching noise are filter topology dependent.

II. Modulation and Foldover Effects

To start the analysis of noise in sampled data systems, we begin with the effects of the idealized sampling operation of Fig. 1 on a signal band limited to less than half the sample

Manuscript received May 6, 1981; revised January 7, 1982.
The author is with Bell Laboratories, Holmdel, NJ 07733.

Fig. 1. Ideal sampler.

rate (f_s). The baseband (η_n) and sideband (η_{sb}) spectral densities of Fig. 2 are equal as a result of ideal impulse sampling [4], [5]. A convention that will prove useful in the analysis to follow is to number the sidebands so as to associate each sideband with the sampling frequency harmonic it is centered about, as in Fig. 2.

For the more general case where the signal bandwidth is greater than $f_s/2$ (such as the output noise of an op-amp in an SCN), aliasing will occur. As an example, assume that the signal to be sampled has a bandwidth (BW_n) of three times the sample rate. The first five sidebands resulting from the sampling operation have been depicted in Fig. 3(a). Referring to the figure, the following are contributors to the frequency band from dc to f_s: the fundamental, ±1, ±2, and the +3 sidebands. The addition of these sidebands is depicted in the stacked structure of Fig. 3(b). This spectral stacking of an undersampled signal results in an output of larger spectral density than the input for the frequency band from dc to $f_s/2$. The noise gain of a track-and-hold circuit is shown in Fig. 4 (and is discussed in Section III).

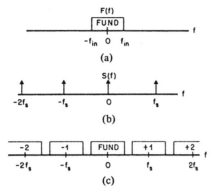

Fig. 2. (a) Input spectrum. (b) Sampling function. (c) Modulated signal spectrum.

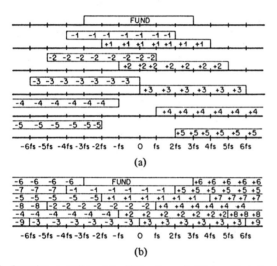

Fig. 3. Impulse sampling (a) sidebands and (b) resultant frequency spectrum.

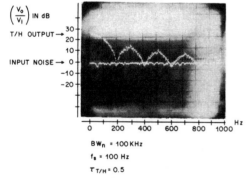

BW$_n$ = 100 KHz
f$_s$ = 100 Hz
$\tau_{T/H}$ = 0.5

Fig. 4. T/H noise response.

For the remainder of this paper, BW_n represents the equivalent noise bandwidth of a noise source, and will be taken to be that bandwidth required to contain the same noise power as the source, but with a uniform spectral density η_n.

If the input is a white noise source, the frequency-shifted sidebands are now uncorrelated with the fundamental or each other; hence, power rather than voltages are added to compute the total output density (η_T). Putting these results into mathematics for $-f_s < f < f_s$,

$$\eta_T = \eta_n + \left[2\left(\frac{BW_n}{f_s}\right) - 1\right]\eta_{sb} \quad (1)$$

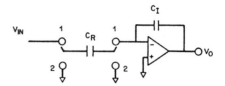

Fig. 5. Integrator topology under study.

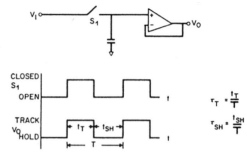

Fig. 6. Track-and-hold circuit.

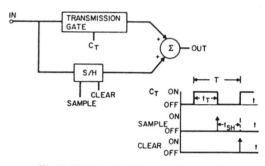

Fig. 7. Track-and-hold equivalent circuit.

where $(2BW_n/f_s) - 1$ is the number of sidebands falling in the frequency range dc $< f < f_s$. To simplify the calculations, BW_n/f_s will be treated as an integer. If a fraction of a sideband folds into band, BW_n/f_s is incremented to the next larger integer to give an upper bound on the in-band noise. Recalling that $\eta_{sb} = \eta_n$ simplifies (1) to

$$\eta_T \leq 2\eta_n\left(\frac{BW_n}{f_s}\right) \quad V^2/Hz. \quad (2)$$

III. NOISE MODEL OF A TRACK-AND-HOLD

One of the basic building blocks commonly used in SCN's is the integrator of Fig. 5. This integrator is a form of a track-and-hold (T/H) circuit. While the track operation is not inherent to all SCN integrator topologies [6]-[8], this property will be included for completeness. The integrator analysis will begin by describing transmission gate (input tracking) and S/H (hold operation) noise responses using the T/H of Fig. 6. These results will be experimentally verified in Section IV, and a complete integrator model will be developed in Section VI.

The T/H can be viewed as two systems in parallel, as in Fig. 7. The transmission gate models the feedthrough operation when the T/H tracks the input. The sample-and-hold (S/H) models the hold operation by sampling at the instant the transmission gate opens, while resetting to zero during the track mode.

A. Track Model

If the input is white noise of bandwidth BW_n, the results of Appendix A can be used to obtain the results below relating the output to the input power density:

$$\eta_{out}(f) \approx \tau_T^2 \eta_n \quad BW_n \leqslant \frac{1}{2} f_s \quad (3a)$$

$$\approx \tau_T \eta_n \quad BW_n \geqslant 10 f_s \quad (3b)$$

where τ_T is the track time/sample period duty cycle. Note that the output spectral density (in the passband) is never greater than that of the input (for a white noise source).

B. Sample-and-Hold Noise Model

An ideal S/H samples instantaneously, and then holds for the remainder of the sampling period. The hold time is the S/H duty cycle times the sample period ($\tau_{SH} \times T$), as in Fig. 7. Using the results of Appendix B for a white noise input,

$$\eta_{out}(f) \leqslant \eta_n \tau_{SH}^2 \operatorname{sinc}^2\left(\frac{\tau_{SH} f}{f_s}\right) \quad BW_n < \frac{1}{2} f_s \quad (4a)$$

$$\eta_{out}(f) \leqslant 2\eta_n \left(\frac{BW_n}{f_s}\right)(\tau_{SH}^2) \operatorname{sinc}^2\left(\frac{\tau_{SH} f}{f_s}\right) \quad BW_n \geqslant \frac{1}{2} f_s. \quad (4b)$$

Thus, foldover effects of ideal sampling followed by a hold operation (that forms the S/H) can produce an in-band response that is significantly larger than the input spectral density.

C. Frequency Domain T/H Noise Model

Since the transmission gate and the S/H operate during non-overlapping time intervals and the noise is white in nature, their power spectra simply add [3], [4]. The track-and-hold operation occupies the entire sampling period:

$$\tau_T + \tau_{SH} = 1.$$

Recasting the transmission gate response in terms of τ_{SH} and combining with the S/H results, the frequency domain model of Fig. 8 can be constructed. Writing $\eta_{out}(f)$ as a function of η_n for wide-band white noise,

$$\eta_{out}(f) = \eta_n \left[\tau_{SH}^2 \operatorname{sinc}^2\left[\frac{\tau_{SH} f}{f_s}\right] + (1 - \tau_{SH})^2\right]$$

$$BW_n < \frac{1}{2} f_s \quad (5)$$

$$\eta_{out}(f) \leqslant \eta_n \left[2\tau_{SH}^2 \left(\frac{BW_n}{f_s}\right) \operatorname{sinc}^2\left[\frac{\tau_{SH} f}{f_s}\right] + 1 - \tau_{SH}\right]$$

$$BW_n \geqslant 10 f_s. \quad (6)$$

To gain physical insight into the above results, let us look at a few limiting cases.

Case 1) $\tau_{SH} = 0$: The tracking gate is always closed, hence no sampling, thus the S/H term drops and $\eta_{out}(f) = \eta_n$.

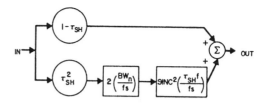

Fig. 8. T/H frequency domain model in terms of S/H duty cycle (τ_{SH}) for $BW_n \geqslant 10 f_s$.

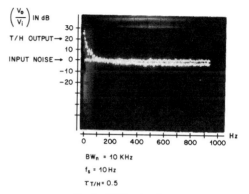

Fig. 9. SCN noise floor.

Case 2) $\tau_{SH} = 1$: Ideal S/H results, and the track term drops, leaving the familiar sinc (x) envelope.

$$\eta_{out}(f) \leqslant 2\eta_n \tau_{SH}^2 \left(\frac{BW_n}{f_s}\right) \operatorname{sinc}^2\left[\frac{\tau_{SH} f}{f_s}\right].$$

Case 3) $\tau_{SH} = 1/2$ and $BW_n > 10 f_s$:

$$\eta_{out}(f) = \eta_n \left[\frac{1}{2}\left(\frac{BW_n}{f_s}\right) \operatorname{sinc}^2\left(\frac{f}{2 f_s}\right) + \frac{1}{2}\right].$$

Case 4: To investigate the out-of-band response, hold f_s and τ_{SH} constant (let $\tau_{SH} = 0.5$ for simplicity), and increase f:

$$\eta_{out}(f) = \eta_n \left(\frac{BW_n}{2 f_s}\right) \operatorname{sinc}^2\left(\frac{f}{2 f_s}\right) + \frac{1}{2} \eta_n.$$

Recalling that $|\operatorname{sinc}(x)| \leqslant 1/|x|$,

$$\leqslant \eta_n \left(\frac{BW_n}{2 f_s}\right)\left(\frac{2 f_s}{f}\right)^2 + \frac{\eta_n}{2}$$

$$\leqslant 2 \eta_n BW_n \frac{f_s}{f^2} + \frac{\eta_n}{2}.$$

As f increases, the envelope of $\eta_{out}(f)$ decreases with the square of f, so for $f \gg f_s$ [see (5) and (6)], the transmission gate term sets the "noise floor" in Figs. 4 and 9 at $\eta_n/2$.

IV. EXPERIMENTAL VERIFICATION OF FOLDOVER THEORY

To aid in the isolation of various noise sources, the test circuit of Fig. 10 was used. To eliminate any effects resulting from finite bandwidth of the op-amps or switches, the noise source was selected to produce band-limited white noise with the cutoff frequency far below f_T of the op-amp. Furthermore, the input noise level was set as far above the noise

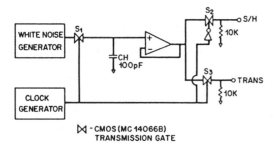

Fig. 10. Transmission gate and S/H test circuit.

TABLE I
(a) TRANSMISSION GATE NOISE RESPONSE VERSUS DUTY CYCLE ($f \ll f_s$) FOR $BW_n = 100$ kHz AND $f_s = 1$ kHz. (b) S/H RESPONSE ($f \ll f_s$) FOR $BW_n = 100$ kHz AND $f_s = 1$ kHz.

(a)

Duty Cycle (in %)	Output Power (W/Hz) referred to input (for $BW_n \geq f_s$)	
	Measured (dB)	Calculated (dB)
100	0	0
95	0	-0.2
61	-1.8	-2.1
50	-2.8	-3.0
35	-4.2	-4.6
18	-7.0	-7.4
5	-13	-13

(b)

Duty Cycle	Output referred to input		First null frequency	
	Measured (dB)	Calculated (dB)	Measured (kHz)	Calculated (kHz)
95	20	22.6	1.0	1.0
80	20	21	1.2	1.4
50	15	17	1.8	2.0
34	12	14	2.8	2.9
21	9	9.4	4.5	4.8
11	4	3.8	8.8	9.1
5	-3	-3	19	20

floor as was practical without circuit limiting. The unswitched (switch latched in closed position) noise signal level serves as the reference for all the results.

A. Track Model Verification

Equation (3) indicates that the track noise contribution is independent of noise bandwidth or sample rate (for $BW_n > 10 f_s$), and is dependent only on the track duty cycle. The calculated and measured track operation results are summarized in Table I(a) with the frequency independence shown in Fig. 11.

B. S/H Noise Model Verification

Using (4b) to predict the S/H response at dc ($\eta_{o\text{max}}$) to a wide-band input,

$$\eta_{o\text{max}} = 2\tau_{SH}^2 \left(\frac{BW_n}{f_s}\right) \eta_n.$$

For $f_s = 1$ kHz, $BW_n = 100$ kHz, and $\tau_{SH} = 0.5$,

$$= 50 \eta_n \text{ (17 dB gain)}.$$

First null occurs at 1 kHz/0.5 = 2 kHz.

The white noise response is shown in Fig. 12, with similarly calculated and measured results summarized in Table I(b).

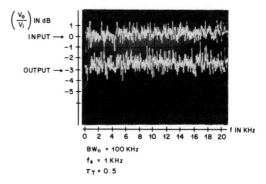

Fig. 11. Transmission gate wide-band input response.

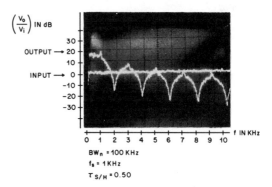

Fig. 12. S/H wide-band input response.

TABLE II
T/H CIRCUIT RESULTS ($\tau_{SH} = 50$ PERCENT) FOR (a) $BW_n = 100$ kHz AND (b) $BW_n = 10$ kHz

(a)

f_s (Hz)	Predicted density (dB)	Measured density (dB)
100k	-0.2	-2
10k	7.4	7.4
1k	17	18
100	27	29

(b)

f_s (Hz)	Predicted Noise (dB)	Measured Noise (dB)
100k	-3	-3.8
10k	0	1.5
1k	7.4	8.5
100	17	17
10	27	26

The S/H response extends below the input noise level because the track operation, with its associated noise contribution, is not present in the S/H.

C. T/H Noise Model Verification

To investigate the foldover term and the orthogonality of the track and S/H operations, the T/H of Fig. 6 has been utilized with a clock duty cycle of 0.5 (50 percent).

Sample Noise Calculation: Calculating the dc spectral density for the reduced sample rate of $f_s = 100$ Hz (measured response in Fig. 4),

$$BW_n = 100 \text{ kHz}, f_s = 100 \text{ Hz}.$$

Using (6),

$$\frac{\eta_{o\text{max}}}{\eta_n} = 2\left(\frac{1}{4}\right)\left(\frac{100 \text{ kHz}}{100 \text{ Hz}}\right) + \frac{1}{2}$$

$$= 500 \text{ (27 dB gain)}.$$

Similar results are to be found in Table II.

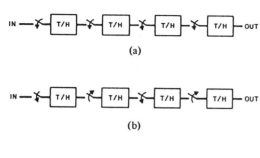
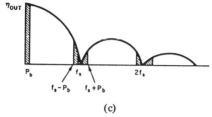

Fig. 13. Cascaded T/H's with in-phase (a) and staggered (b) phase clocking. The resampled wide-band noise spectrum for the staggered clock case is shown in (c).

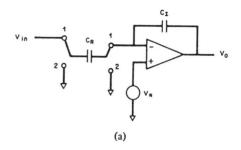
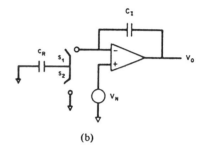

Fig. 14. (a) Charge accumulator and (b) simplified accumulator with V_{in} grounded.

V. EXTENSION TO CASCADED STAGES

There are two extreme cases when cascading T/H stages. One is when all the T/H's have the same timing (in-phase clocks) as in Fig. 13(a). The other extreme is when one T/H tracks, and the T/H before and after it are in the hold state (staggered clocks) as in Fig. 13(b). While parts of an SCN might contain both clocking schemes, SCN realization considerations make the staggered clock case the most common [6], [7], [11].

In one extreme, all the T/H stages are driven by the same clock phases. Because the significant foldover effects are present only in the hold phase (assuming a memoryless track phase), the noise at the T/H input is simply the sum of the unsampled noise feeding through to the stage of interest. When the T/H changes to the hold phase, the input noise is sampled and folded down to baseband. The resulting output spectra are described by (5) and (6).

For the other extreme (staggered clocks), each successive T/H samples the held output of the previous stage. This sampling of a held output results in $\tau_{SH} = 1$ and the output spectrum of Fig. 13(c) where P_b is the passband edge. It is easily shown that the sin x/x envelope provides effective anti-alias filtering of the folded noise (increasing with larger f_s/P_b ratios) so that resampling the held T/H output minimally increases the in-band noise. As staggered T/H's are cascaded, the output noise of a given stage (in the hold phase) is closely approximated by the resampled noise of the previous stages added to the sampled wide-band noise of the stage the present T/H is sampling.

In summary, since most SCN's use staggered clocks, τ_{SH} will be taken to be unity for the rest of the paper, and the effects of resampling of noise will be ignored.

VI. SCN INTEGRATORS AND CHARGE ACCUMULATORS

In this section, a model of an SCN integrator is developed. The T/H results will then be combined with these results in Section VII where a simple SCN is analyzed.

The topology of the charge accumulator (with the 1/f and flatband op-amp noise modeled by V_n) of Fig. 14 is representative of a basic structure used in SCN filters [6]-[8], so a careful analysis is in order.

A. Charge Accumulator Model

To aid in visualization, the accumulator is redrawn in Fig. 14(b) with V_{in} grounded. Assume as initial conditions that $q_{CI} = q_{CR} = 0$ and that the op-amp voltage gain is infinite. After switch S_1 has closed, the feedback action of C_I will force the output to

$$V_o = \left(1 + \frac{C_R}{C_I}\right) V_n$$

(while driving the inverting input to equal V_n) and to track V_n with gain thereafter. The transient resulting from C_R charging to V_n does not trap charge on C_I because only the trapped charge from previous sample periods will remain if the amplitude of V_n returns to zero at the instant S_1 opens. The additional charge trapped on C_I at the instant S_1 opens (at time t1) is

$$\Delta q_{CI}(t1) = q_{CR} = C_R v_n(t1)$$

with

$$\Delta V_{C_I}(t1) = \frac{C_R}{C_I} V_n(t1).$$

With S_2 closed and S_1 open, q_{CR} is bled off to ground, and V_o tracks V_n with unity gain and a dc offset equal to the voltage stored on C_I. When S_2 opens and S_1 closes, the inverting terminal is initially pulled to ground, and the op-amp then recharges C_R to V_n, repeating the above procedure. Note that the only time charge is trapped on C_I is when S_1 opens, resulting in an accumulated output offset (V_{net}) of

$$V_{net} = \sum_{i=1}^{i=N} \frac{C_R}{C_I} V_n(iT) \qquad (7)$$

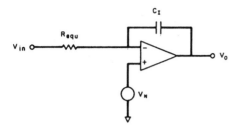

Fig. 15. SCN integrator noise model for $C_R/C_I \ll 1$ and $R_{equ} = 1/C_R f_s$.

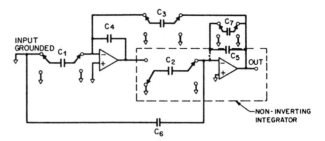

Fig. 16. HPN noise test circuit.

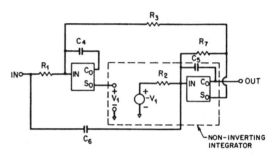

Fig. 17. Noise simulation model for HPN SCN.

where $V_n(iT)$ is the noise voltage at the instant S_1 opens, T is the complete S_1, S_2 period, and N is the number of complete clock periods since C_I was discharged. The output when S_1 is closed is

$$V_o = \left(1 + \frac{C_R}{C_I}\right) V_n(t) + V_{net}(N-1). \qquad (8)$$

When S_2 is closed,

$$V_o = V_n(t) + V_{net}(N) \qquad (9)$$

where the index of V_{net} has been incremented because additional charge has been trapped on C_I when S_1 opened.

If there are N sample periods per clearing of C_I (as in some low-loss integration techniques), and V_n is Gaussian white noise, probability theory indicates that $V_{net}(N)$ will have a standard deviation $\sqrt{N}$ times larger than $(C_R/C_I) V_n$. In other words, the noise power of V_{net} after N cycles will be N times larger than the incremental noise power stored on C_I in any given sample cycle.

B. SCN Filter Approximations

A significant simplification can be made for those filter stages with pole frequencies much less than the sample rate. These filters usually have $C_R/C_I \ll 1$. Under these conditions, (7) shows that the individual contributions to V_{net} are less than the referred to input noise V_n by C_R/C_I. In addition, the number of terms in (7) that significantly contribute to the integrator's output noise is limited by dc discharge paths from the amplifier output to its input. These discharge paths are used to provide Q damping and dc stability of the filter. In most practical SCN's, these conditions result in the input referred noise of the op-amp (V_n) being much larger than V_{net}, so the V_{net} term will be dropped.

To model charge transfer uncertainty and the foldover effects of the wide-band noise of the switches associated with C_R, an equivalent resistance of value [9]

$$R_{equ} = \frac{1}{C_R f_s}$$

is used in the continuous-time equivalent circuit of Fig. 15. These approximations hold when the filter passband is much less than the sample rate. For the noise simulations of the next section, the SCN integrator will be replaced by its continuous-time counterpart. The output noise of a particular SCN will be worked out in detail in Section VII to demonstrate how to use these results.

VII. NOISE MODEL AND CALCULATIONS FOR SCN's

This section will develop a model suitable for SPICE noise simulations of switched capacitor networks. Experimental results in support of these results will also be given.

A. SCN Topology Model

To simulate the capacitor switching noise (kT/C), replace the switched capacitors of the HPN in Fig. 16 with resistors of value

$$R = \frac{1}{C f_s}$$

as per Fig. 17.

Next replace the op-amps with the blocks shown in Fig. 17. The switched output (S_o) is connected to all the switched capacitors connected to the output of the op-amps. The continuous output (C_o) is connected to the paths that are not switched. To model the noninverting integrator of Fig. 16, use a voltage-controlled voltage source of gain = -1 to provide the polarity reversal (as in Fig. 17). The filter topology has now been modeled; next the op-amp model will be developed.

B. Op-Amp Model

To model the op-amp of Fig. 18 in an SCN environment, the sampled data effects on the output noise of the op-amp must be considered. This noise consists of two parts: wideband (flat, white noise), and $1/f$ (flicker) noise. The voltage sources $V_{1/f}$ and V_{n1} are used to model the unsampled $1/f$ and flat-band noise, respectively. Since the foldover effects are added to the unsampled output noise, the noise sources kV_{n11} and V_{f11} have been placed in series with the continuous output to yield S_o. Referring to Appendix C to determine the op-amp BW_n, the foldover factor (k) is derived by simplifying (6) under the constraints that $f \ll f_s$, so the $\sin x/x$ term

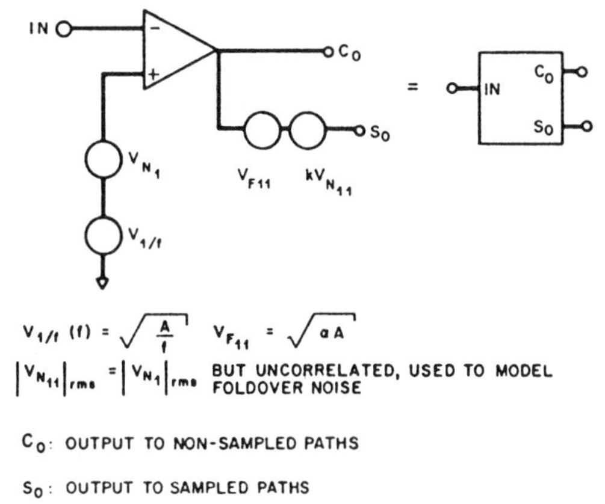

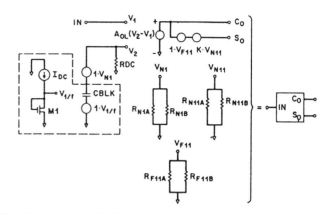

Fig. 18. Frequency domain SCN op-amp noise simulation model.

Fig. 19. Complete SPICE op-amp noise model. The boxed circuit is the $1/f$ source, and the resistors are uncorrelated white noise sources.

approaches unity and $\tau_{SH} = 1$. Then separate the result into terms related to the C_o and S_o outputs, respectively:

$$\eta_{\text{out}}(f) \leqslant \eta_n + \eta_n \left[\frac{2BW_n}{f_s} - 1\right]$$

with the bracketed term representing the foldover effects represented by kV_{n11}. Similarly accounting for the $1/f$ contribution (Appendix C),

$$\eta_{1/f}\text{ out }(f) \leqslant \frac{A}{f} + (\alpha A).$$

Because programs such as SPICE work with noise voltages rather than powers, the noise voltage gain through sampling is obtained by taking the square root of the bracketed terms to yield

$$k = \sqrt{\frac{2BW_n}{f_s} - 1}$$

and the $1/f$ term of $\sqrt{\alpha A}$.

An alternate method of calculating "k" is as follows. First, calculate the total output noise power (P_T) of the amplifier using SPICE or other means. The result is doubled to account for positive and negative sidebands' contributing power, and is divided by the sample rate to yield an average power density. Putting this procedure into mathematics,

$$k = \sqrt{\frac{2P_T}{f_s \eta_{FB}} - 1}$$

where η_{FB} is the presampled op-amp flat-band power density.

The next two subsections will outline in detail how to use SPICE to model the op-amp noise.

C. Flicker Noise Model

Because SPICE lacks the option of user-defined functional expressions, a device model must be used to simulate the unsampled $1/f$ noise component. To conveniently generate the flicker noise, the boxed circuit of Fig. 19 is used. The device is diode-connected for ease of biasing to a predetermined drain current by I_{dc}. To assure that the device is in the saturation region, choose M1 as an enhancement-type device. The noise voltage as seen at the drain is

$$V_n = \sqrt{\frac{8kT}{3gm} + \frac{KF}{fWL_{\text{eff}}}}.$$

To decouple the noise sources so that $1/f$ and wide-band noise effects can be studied separately, select the bias current and device W/L ratio to assure that the gm term contributes much less noise than the input-referred op-amp flat-band noise. With the device parameters selected, adjust KF to match the measured op-amp input-referred $1/f$ noise.

Because the flicker noise is to be injected into the noninverting terminal of a high gain op-amp, the dc bias voltage at the drain must be blocked. To avoid loading the drain with the dc blocking circuit (RDC, $CBLK$), a voltage-controlled voltage source is used as an ideal unity gain buffer, as in Fig. 19.

The next subsection will show how to model the sampled $1/f$ noise.

D. Flat-Band and Foldover Model

The flat-band, folded flat-band, and folded $1/f$ (Appendix C) can be modeled as white noise sources, which are easily simulated by resistors in SPICE.

To model the flat-band noise, the resistor value is selected to match the measured input referred flat-band noise. To satisfy the nodal requirements of SPICE, two resistors (R_{n1A} and R_{n1B} of Fig. 19), each twice the value of the input-referred noise resistance, are paralleled so that all nodes have at least two components connected and a dc path to ground. The source V_{n1} injects the unsampled flat-band noise into the op-amp, with V_{n11} modeling the wide-band foldover effects and V_{f11} the $1/f$ folded noise. Separate resistor sets are used to drive V_{n1}, V_{n11}, and V_{f11} so that the sources will be mutually uncorrelated.

The appropriate folded $1/f$ resistor values are

$$R_{F11A} = R_{F11B} = 2\frac{\alpha A}{4kT}.$$

E. Model Limitations

This model is accurate for $f \ll f_s$, so the $\sin x/x$ frequency shaping will not appreciably affect the in-band noise. This

TABLE III
MONOLITHIC FILTER RESULTS. NOISE VOLTAGES ARE rms VALUES INTEGRATED OVER THE FREQUENCY RANGE FROM 1 Hz TO 10 kHz.

FILTER	1/f	FLAT	CAPACITOR	TOTAL NOISE	dBrnc*	MEASURED dBrnc
HPN $f_s = 8kHz$	4uv	6.5uv	9.0uv[1]	12uv	-6.5	-6
HPN $f_s = 64kHz$	4uv	3.9uv	4.1uv[2]	6.8uv	-11	-8
LPF $f_s = 128kHz$	12uv	25uv	45uv[3]	53uv	6.7	8.5
LPF reduced caps.	12uv	25uv	100uv[4]	110uv	13	15
LPF reduced caps. & input dev.	27uv	25uv	100uv[4]	110uv	13	16

* 0 dBrn (24.5uv)
[1] $C_{unit} = 0.6pF$
[2] $C_{unit} = 0.3pF$
[3] $C_{unit} = 1.7pF$
[4] $C_{unit} = 0.3pF$

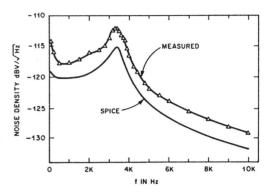

Fig. 20. Measured versus calculated 1.7 pF unit capacitor LPF noise response.

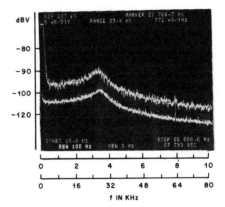

Fig. 21. Measured noise response of the 1.7 pF unit capacitor LPF for $f_s = 128$ kHz (top trace) and for $f_s = 1024$ kHz (bottom trace). All analyzer settings are the same for both traces except that the lower trace stop frequency (80 kHz) is eight times that of the top trace (10 kHz).

constraint is not severe since most SCN's are designed with the sample rate much greater than the passband frequency. Further assumptions made to simplify the calculations are that the hold duty cycle of the switched paths (τ_{SH}) is unity and that the filter is run at a single sample rate. If stages of differing sample rates are cascaded, the sampled noise spectra of previous stages must be analyzed to account for significant foldover contributions to the stage under consideration.

F. Calculation and Measurement Results

Using models like the above, SPICE simulations were used to calculate the output noise for 60 Hz HPN filters sampled at 8 and 64 kHz, and several 128 kHz sample rate fifth-order elliptic low-pass filters with a cutoff frequency of 3.4 kHz [8]. The calculated C-message weighted (dBrnc) noise results are summarized in Table III, and are compared to results obtained from monolithic realizations of these circuits. The op-amp noise was measured to be 50 nV/$\sqrt{Hz}$ at 1 kHz, with a flat-band noise component of 35 nV/$\sqrt{Hz}$ and a noise bandwidth of 2 MHz.

Fig. 16 shows an actual HPN (sampling at 8 kHz) switched capacitor filter [8], and its complete noise model is shown in Fig. 17. Calculating k for the filter with op-amps having a noise bandwidth (see Appendix C) of 2 MHz,

$$k = \sqrt{2 \frac{BW_n}{f_s} - 1} = \sqrt{2 \frac{2 \text{ MHz}}{8 \text{ kHz}} - 1} = 22.3.$$

Referring to Table V indicates that $\alpha \approx 2.1$ for this sample rate.

The dominant noise term of the HPN is the capacitor switching noise. If the filter is followed by an 8 kHz S/H, the wideband noise of the second amplifier will be boosted by the foldover factor (22.3) to become the dominant contributor. Similarly, the dominant noise source in the LPF's is the capacitor switching noise.

The measured and predicted noise spectra are compared in Fig. 20 for the 1.7 pF minimum capacitor size LPF. If wideband noise sources are the dominant terms of the LPF, the noise density should drop by 9 dB for a factor of 8 increase in sample rate. As a test, the noise responses for $f_s = 128$ and 1024 kHz of this LPF are compared in Fig. 21. All the analyzer settings are the same for both traces, with the exception that the frequency sweep was changed from 10 kHz up to 80 kHz (for $f_s = 1024$ kHz) so the relative densities can be directly compared. The approximate 10 dB drop in the passband noise density clearly indicates that 1/f noise is definitely not the dominant noise source of these filters.

VIII. CONCLUSION

The results presented show that the topology of an SCN determines the significance of the relative contributions of amplifier flat-band and 1/f, and capacitor switching noise to the overall filter noise. For the test chip filter designs in our process, the dominant noise source was found to be capacitor switching noise, followed by op-amp wide-band noise, and lastly, op-amp 1/f noise. A noise model and a simulation technique have been developed to allow existing simulation programs (such as SPICE) to be used in accurate SCN noise calculations.

APPENDIX A

Transmission Gate Noise Model

The transmission gate simply modulates the input signal $\{x(t)\}$ by a pulse train, as shown in Fig. 22. The Fourier

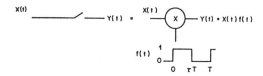

Fig. 22. Transmission gate model.

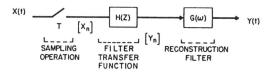

Fig. 23. General topology of a sampled data system.

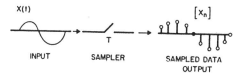

Fig. 24. Ideal sampling switch operation.

TABLE IV
Transmission Gate Noise Response as a Function of Duty Cycle and BW_n/f_s

DUTY CYCLE τ	η_{out}/η_n as a function of $\frac{BW_n}{f_s}$				
	100	10	6	1	½*
1.00	1.00	1.00	1.00	1.00	1.00
0.95	0.95	0.94	0.93	0.91	0.90
0.80	0.80	0.79	0.78	0.71	0.64
0.60	0.60	0.59	0.59	0.54	0.37
0.50	0.50	0.49	0.48	0.45	0.25
0.35	0.35	0.34	0.33	0.28	0.12
0.20	0.20	0.19	0.18	0.11	0.04
0.10	0.10	0.090	0.086	0.029	0.010
0.05	0.049	0.040	0.029	0.007	0.003
0	0	0	0	0	0

*(foldover effects absent)

series of the modulating pulse is

$$f(t) = \tau + \frac{2}{\pi} \sum_{i=1}^{\infty} \frac{(-1)^i}{i} \sin(i\pi\tau) \cos\left(\frac{i 2\pi t}{T}\right) \quad (A1)$$

where τ is the duty cycle and T is the sample period. Working out the details for a white noise input of bandwidth BW_n,

$$\eta_{out}(f) = \eta_n \left[\tau^2 + \frac{1}{2} \left(\frac{2}{\pi}\right)^2 \sum_{i=1}^{h} \left[\frac{1}{i} \sin(i\pi\tau)\right]^2 \right] \quad (A2)$$

where $h = BW_n/f_s$ (number of positive sidebands falling in band).

Since (A2) is not in a convenient form for h, $\eta_{out}(f)$ has been tabulated in Table IV for various BW_n/f_s and τ. An important approximation can be made for $h \geq 10$:

$$\eta_{out}(f) \approx \tau \eta_n \quad \text{for } \frac{BW_n}{f_s} \geq 10, \quad (A3)$$

while at the other extreme, $h = 0$ (sin² terms are not present),

$$\eta_{out}(f) \approx \tau^2 \eta_n \quad \text{for } 2BW_n \leq f_s. \quad (A4)$$

APPENDIX B
A Sample-and-Hold Noise Model

A general topology for a discrete-time data system is shown in Fig. 23. The sampling function is shown in Fig. 24. The S/H operation consists of ideal sampling, followed by filtering with $H(z) = 1$, feeding into a hold operation. The hold reconstructs the impulse train into a series of levels which constitute the familiar held output response.

Calculation of $\{x_n\}$ Spectrum

Taking the Fourier transform of $\{x_n\}$ [10], we find

$$X(e^{j2\pi fT}) = \frac{1}{T} \sum_{k=-\infty}^{\infty} X^a \left[2\pi f + \frac{2\pi k}{T}\right] \quad (B1)$$

where X^a is the Fourier transform of $x(t)$, which is periodic about multiples of the sampling frequency.

Calculation of the S/H Frequency Response

Referring to the timing of Fig. 25, we can determine the limits of integration:

$$G(2\pi f) = \int_{t=-\infty}^{\infty} g(t) e^{-j2\pi ft} \, dt = \int_{t=0}^{h} e^{-j2\pi ft} \, dt$$

$$= h e^{-j2\pi f(h/2)} \operatorname{sinc}(fh)$$

where

$$\operatorname{sinc}(x) = \frac{\sin(\pi x)}{\pi x}$$

h = (hold duty cycle) (sample period)

$$= \tau T = \frac{\tau}{f_s}.$$

Substitution yields

$$G(2\pi f) = \tau T \operatorname{sinc}\left[\tau \left(\frac{f}{f_s}\right)\right] e^{-j2\pi f(h/2)}. \quad (B2)$$

When dealing with noise, power is usually the quantity of interest; calculating the output spectrum,

$$|Y(2\pi f)|^2 = |X(e^{j2\pi fT}) G(2\pi f)|^2 = |X(e^{j2\pi fT})|^2 |G(2\pi f)|^2.$$

Calculating $|X(e^{j2\pi fT})|^2$,

$$|X(e^{j2\pi fT})|^2 = \frac{1}{T^2}\left|\sum_{k=-\infty}^{\infty} X^a\left[2\pi f + \frac{2\pi k}{T}\right]\right|^2.$$

For the case at hand, $x(t)$ is white noise; hence, the aliases are uncorrelated, with the result that power, rather than voltage, terms are added. Using (2),

$$|X(e^{j2\pi fT})|^2 = \frac{2}{T^2}\eta_n\left(\frac{BW_n}{f_s}\right) \quad \text{for } f_s \leqslant 2BW_n \quad \text{(B3a)}$$

$$= \eta_n \frac{1}{T^2} \quad \text{for } f_s \geqslant 2BW_n, \quad \text{(B3b)}$$

and for the S/H,

$$|G(f)|^2 = \tau^2 T^2 \operatorname{sinc}^2\left[\tau\left(\frac{f}{f_s}\right)\right]. \quad \text{(B4)}$$

Combining (B3) and (4) yields

$$|Y(f)|^2 = 2\eta_n\left(\frac{BW_n}{f_s}\right)\tau^2 \operatorname{sinc}^2\left[\tau\left(\frac{f}{f_s}\right)\right] \quad f_s \leqslant 2BW_n \quad \text{(B5a)}$$

and

$$|Y(f)|^2 = \eta_n \tau^2 \operatorname{sinc}^2\left[\tau\left(\frac{f}{f_s}\right)\right] \quad f_s \geqslant 2BW_n. \quad \text{(B5b)}$$

The usual application of sampled data systems is the processing of information in the passband ($f \leqslant f_s/2$); accordingly, the in-band signal-to-noise ratio is of interest. Since signal-to-noise is given by

$$\frac{S}{N} = \frac{|\text{input }(2\pi f)|^2 |G(2\pi f)|^2}{|X(e^{j2\pi fT})|^2 |G(2\pi f)|^2} = \frac{|\text{input }(2\pi f)|^2}{|X(e^{j2\pi fT})|^2},$$

it is independent of the reconstruction filter in the passband for white noise input, but does depend on the undersampling of noise.

Appendix C

Foldover Effects and Equivalent Noise Bandwidth

This analysis will treat the op-amp $1/f$ and wide-band noise separately to clearly show the different effects of undersampling of these noise sources and to point out useful approximations.

Wide-Band Noise

Fig. 26 depicts a characteristic voltage-follower output noise spectrum (of the wide-band component only) that has been divided into frequency bands of width f_s. Settling time constraints of SCN's usually restrict f_s to be much less than the op-amp unity gain frequency. Additionally, the switch bandwidths are usually set much wider than that of the op-amp. With the wide-bandwidth switches, effectively all the output noise of the op-amp is folded back in to the baseband from dc

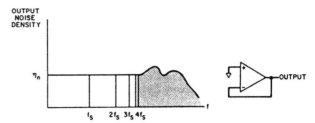

Fig. 26. Voltage-follower output noise spectrum with test configuration.

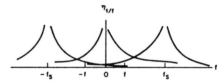

Fig. 27. Fundamental and first two sidebands of a sampled $1/f$ noise source.

to f_s. The total op-amp output noise is calculated by adding up the power in each frequency band of width f_s.

The equivalent noise bandwidth (BW_n) of an op-amp will be taken to be that bandwidth required to contain the same noise power as the op-amp, but with a uniform spectral density η_n.

$1/f$ Foldover Effects

Idealized sampling of a $1/f$ noise source produces the spectrum of Fig. 27 where only the first sidebands have been shown for simplicity. In the analysis that follows, it is assumed that the noise spectral density follows an amplitude envelope of the form A/f, and that the sidebands are mutually uncorrelated so that power, rather that voltage, is summed.

Evaluating the foldover effects in the baseband from dc to f_s,

$$\eta(f) = A\left[\frac{1}{f} + \sum_{i=1}^{N}\left(\frac{1}{if_s - f} + \frac{1}{if_s + f}\right)\right] \quad \text{(C1)}$$

where N is the number of sidebands folding into the baseband.

To understand the foldover effects described in (C1), let us assume that the voltage-follower op-amp attenuates the $1/f$ noise for frequencies above 20 MHz, $A = 1000$ for computation ease, and that the frequency range of interest is the familiar telephone voiceband from 300 to 4 kHz. Fig. 28 compares the presampled and postsampled spectra for several common clock rates. An important aspect of Fig. 28 is that all the process dependence is lumped into the constant A, so these curves are simply multiplied by the same scale factor to fit any process. Evaluating just the foldover terms of (C1) (and calling that sum α) indicates that the $1/f$ foldover effects can be closely approximated by adding a constant to the pre-sampled spectrum. Also note that for typical sample rates of 64 kHz or higher, the foldover contribution is less than 20 percent of the $1/f$ baseband density at 1 kHz. A simplifying approximation is to neglect the $1/f$ foldover term entirely for a sample rate of 100 kHz or higher. For lower sample rates,

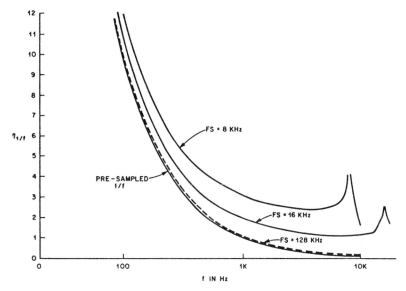

Fig. 28. Pre- and postsampled $1/f$ spectra for several sample rates. For these curves, $V^2_{1/f} = A/f$ with $A = 1000$.

TABLE V
$1/f$ Foldover Factor in Units of Hz^{-1}

f (Hz)	$f_s = $ 8kHz	16kHz	32kHz	64kHz	128kHz	256kHz
300	2.1	0.96	0.44	0.20	0.09	0.04
1000	2.1	0.96	0.44	0.20	0.09	0.04
2000	2.1	0.97	0.44	0.20	0.09	0.04
3000	2.2	0.97	0.44	0.20	0.09	0.04
3500	2.2	0.97	0.44	0.20	0.09	0.04
4000	2.2	0.97	0.44	0.20	0.09	0.04

place a white noise source in series with the sampled op-amp output (Fig. 16) of magnitude αA as summarized in Table V.

ACKNOWLEDGMENT

The author is grateful to P. E. Fleischer, A. Ganesan, D. G. Marsh, and K. R. Laker for their helpful discussions and ideas. Special thanks go to A. A. Schwarz for the excellent layout and laboratory work performed.

REFERENCES

[1] C.-A. Gobet and A. Knob, "Noise analysis of switched capacitor networks," in *ISCAS Proc.*, Apr. 1981, pp. 856–859.
[2] B. Furrer and W. Guggenbuhl, "Noise analysis of sampled-data circuits," in *ISCAS Proc.*, Apr. 1981, pp. 860–863.
[3] M. L. Liou and Y-L. Kuo, "Exact analysis of switched capacitor circuits with arbitrary inputs," *IEEE Trans. Circuits Syst.*, vol. CAS-26, pp. 213–223, Apr. 1979.
[4] A. B. Carlson, *Communication Systems*. New York: McGraw-Hill, 1968.
[5] R. A. Gable and R. A. Roberts, *Signals and Linear Systems*. New York: Wiley, 1973.
[6] G. M. Jacobs, "Practical design considerations for MOS switched capacitor ladder filters," Electron. Res. Lab., Univ. California, Berkeley, Memo. UCB/ERL M77/69, Nov. 1977.
[7] D. J. Allstot, "MOS switched capacitor ladder filters," Ph.D. dissertation, Electron. Res. Lab., Univ. California, Berkeley, Memo. UCB/ERL M79/30, May 1979.
[8] P. E. Fleischer and K. R. Laker, "A family of active switched capacitor biquad building blocks," *Bell Syst. Tech. J.*, vol. 58, pp. 2235–2269, Dec. 1979.
[9] C.-A. Gobet and A. Knob, "Noise generated in switched capacitor networks," *IEE Electron. Lett.*, vol. 16, pp. 734–735, Sept. 1980.
[10] A. Peled and B. Liu, *Digital Signal Processing*. New York: Wiley, 1976.
[11] P. E. Fleischer, A. Ganesan, and K. R. Laker, "Effects of finite op amp gain and band-width on switched capacitor filters," presented at the IEEE Int. Symp. Circuits and Syst., Apr. 1981.

Noise Analysis of Switched Capacitor Networks

CLAUDE-ALAIN GOBET AND ALEXANDER KNOB, MEMBER, IEEE

Abstract —Noise generated in switched capacitor (SC) networks has its origin in the thermal fluctuations of charged particles in the channels of the MOS switch transistors on one hand, in the operational amplifiers on the other hand. Using a SC integrator as vehicle, it is shown that the output noise spectrum consists in general of a broad-band component due to a continuous-time noise signal and of a narrow-band contribution predominating in the baseband of the SC network resulting from a sampled-data noise signal. The ratio of undersampling is introduced and it is shown that the latter noise contribution can be evaluated by sampled-data techniques using the z-transform transfer function. These results are applied to the SC integrator and excellent concordance to the measurements made on a laboratory model is established.

I. INTRODUCTION

A SERIOUS PROBLEM arises when designing low-noise switched capacitor (SC) circuits since they produce in general higher noise levels than their RC-active counterparts. There is hence a need to understand the phenomena which lead to generation of noise in SC networks.

Although a great variety of SC analysis methods was presented in the last few years, just in a few papers [6], [7] the problem of noise has been mentioned. This is essentially due to the fact that for most methods SC networks without resistive components (i.e., transients) are assumed. For such networks, the aliasing of higher noise spectral components would result in a infinite power in the baseband. Also the fact that the bandwidth of noise signals present in SC networks exceeds in most practical cases the sampling frequency by several orders of magnitude implies that the required approach of analysis be essentially different from the case where signals satisfying the Nyquist criterion are present. The first noise calculations on SC networks have been performed by Allstot [1] for a parallel SC integrator. The output noise was attributed to the thermal noise associated with the on-resistance of the MOS switch transistors and the equivalent input noise of the MOS op amp. This latter was assumed to have a flicker component dominating in the low-frequency range. Recently, the same SC configuration has been reinvestigated independently by the authors [3] and some more accurate evaluations have been set up, without considering however the flicker noise component of the op amp. Similar results have been obtained by Furrer and Guggenbühl [4] for a SC inverting amplifier using the correlation technique and by Maloberti *et al.* [5] for a bilinear SC integrator. In the latter paper however, just the op amp's noise has been considered.

In this paper, the SC integrator investigated in [3] will be used as a vehicle to set up more general statements on SC noise analysis. In Section II it will be outlined that the output noise of a SC network can be split into a direct broad-band component which is in most practical cases negligible for the interesting frequency range up to the Nyquist rate, and into a sampled-and-held contribution due to the sampling of broad-band components. While the evaluation of the "direct noise" contribution (due to noise sources with direct output coupling in at least one phase) can be carried out by classical analysis techniques, for the determination of the latter component sampled-data techniques can be used. It will be shown furthermore, that the evaluation of the "sampled noise" component, predominating up to the Nyquist rate, can be made by considering the SC network as transientless (without resistors and op amp poles). The band limitation of the white noise sources, caused by the transients present in SC networks, is implicitely contained in the "ratio of undersampling" of the sampled broad-band noise spectrum (Section III). This method can be of great use in SC network noise analysis: it permits the noise analysis of SC networks by "classical" analysis programs which do not take into account the transients present in SC networks. Finally, the noise spectral density of a SC integrator is measured (Section IV), and its good correspondence with the analytical results confirms the noise model established.

II. SC INTEGRATOR'S NOISE EQUIVALENT

The simple parallel switched integrator, as it has been implemented in the first SC realizations, is shown in Fig. 1, with the gates of the MOS transistors (MOST) controlled by two nonoverlapping clocks Φ_1, Φ_2 of period T and duty cycle Δ/T. We propose to evaluate the spectral distribution of the output noise voltage $S_n(\omega)$ for a shunted input. The output noise is assumed to have two origins: the thermal noise created in the resistance of the MOS switches and the noise generated in the operational amplifier. Since the different noise sources are uncorrelated, their contribution to the output noise spectrum can be evaluated separately.

Manuscript received November 12, 1980; revised May 10, 1982.
C.-A. Gobet is with the Electronics Laboratory of the Swiss Federal Institute of Technology Lausanne, av. de Cour 33, 1007 Lausanne, Switzerland.
A. Knob is with the Institut de Recherches Robert Bosch S.A., post-box 18, 1027 Lonay, Switzerland.

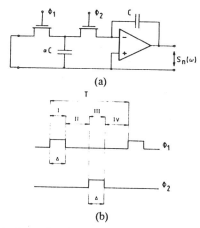

Fig. 1. Switched capacitor integrator with (a) input shunted, controlled by (b) two nonoverlapping clock signals Φ_1 and Φ_2 (transistors conducting for high level) of period T, duty cycle Δ/T and corresponding time-slots I–IV.

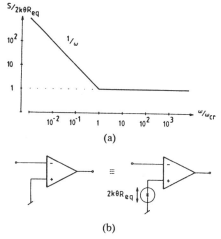

Fig. 2. Typical plot of (1) equivalent MOS op amp input noise spectral density and (b) noise model (white noise voltage source with noiseless op amp) for grounded op amp with one pole rolloff characteristic (R_{eq} is the equivalent input noise resistance, k is the Boltzmann's constant, θ is the absolute temperature).

Fig. 3. Noise model of the MOSFET switch transistor with $R_\Phi(t) = R_{on}$ (Φ high) and $R_\Phi(t) = R_{off}$ (Φ low), the on and off-resistances of the switch transistors, respectively.

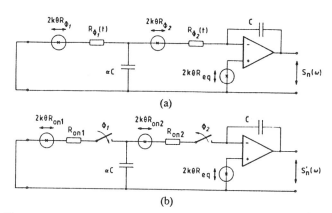

Fig. 4. SC integrator's exact time-variant noise equivalent (a) and (b) simplified noise model (noise contributions of the MOSFET transistors in the off-state neglected).

A. Noise Model for Op Amp and MOS Switch Transistors

The equivalent input noise spectral density of a MOS operational amplifier is frequency dependent as illustrated in Fig. 2(a). The power spectral density at low frequencies is dominated by the flicker component which decreases as $1/\omega$, determined by the area of the input transistors. The spectral density at higher frequencies is constant (typically $0,1 \mu V/\sqrt{Hz}$ for C-MOS low-power devices) and depends on the bias current of the input stage. The latter noise contribution can be represented as the thermal noise $2k\theta R_{eq}$ (bilateral representation) produced in a pseudoresistance at one input of the op amp (with k the Boltzmann's constant and θ the absolute temperature).

Depending on the passband frequency and the sampling rate of the filter, one or both components may be important in determining the dynamic range at the filter's output. However, in this paper, for the sake of simplicity, only the white noise component will be considered. This assumption is realistic in many practical cases, since the flicker contribution is submerged, as it will be stated later, by the aliased broadband components. It will be assumed furthermore that the op amp has a low-frequency gain of A_0 with a one pole rolloff

$$|A(\omega)| = \frac{A_0}{\sqrt{1+(A_0\omega/\omega_u)^2}} \quad (1)$$

where ω_u is the unity gain bandwidth. The noise model adopted for the op amp is shown in Fig. 2(b).

The noise contributions from the MOS switch transistors also need to be considered. The derivation of their noise model is straightforward and yields a series connected noiseless resistor R_Φ with a noise voltage source of uniform spectral density $2k\theta R_\Phi$ (see Fig. 3). The noise model depends thus on the clock signal Φ, with $R_\Phi = R_{on}$ (Φ high) and $R_\Phi = R_{off}$ (Φ low), the on and off resistances of the MOS switch transistors, respectively.

B. Complete Noise Model of the SC Integrator

By means of the results derived in the foregoing section, the noise equivalent of the SC integrator can be established in a straightforward manner, as shown in Fig. 4(a). Note that the noise equivalent is a time-variant four-phase network, depending on the time intervals I, II, III, and IV of Fig. 1. The noise signals issued from the white noise voltage sources are lowpass filtered by the different RC time constants and/or by the op amp's rolloff characteristic. Since in current MOS technologies the on and off resistances of the MOS switches differ up to ten orders of magnitude, the noise power due to the switch transistor in the off state is concentrated at extremely low frequencies. This contribution can be considered as a slowly varying offset voltage and will henceforth be ignored. Each nonconducting MOS switch transistor is therefore replaced by an open circuit, the four-phase noise equivalent is hence reduced to a resistive SC network with internal noise sources, as illustrated in Fig. 4(b). Note that the output noise spectral density $S_n'(\omega)$ consists of two components: a broadband noise contribution (due to the sources $2k\theta R_{on2}$ and $2k\theta R_{eq}$ while Φ_2 is active and only to the latter when

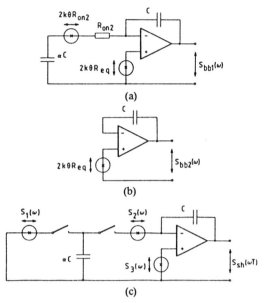

Fig. 5. Decomposition of simplified noise equivalent of Fig. 4(b) into two time-invariant subcircuits, valid for the intervals (a) Φ_2 high and (b) Φ_2 low, and (c) a two-phase transientless sampled-data network with band-limited noise sources.

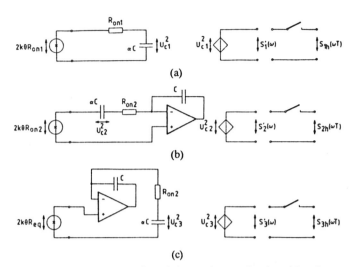

Fig. 6. Possible representation of the continuous-time broad-band contributions $S'_j(\omega)$ (for $j = 1, 2, 3$) of the corresponding noise source $S_j(\omega)$ on the capacitor αC and theirs sampled spectra $S_{jh}(\omega T)$.

Φ_2 is inactive) and a sampled-and-held noise component due to all three noise sources at the end of phase Φ_2.

In SC networks, all time constants must be much smaller than the clock period, in order to enable a total charge transfer. This condition implies that the noise cutoff (determined by the RC time constants and the unity gain bandwidth of the op amp) is situated at higher frequencies than the sampling rate. These broad-band noise components are in consequence undersampled. This means in terms of the noise autocorrelation function that the correlation time constant is much smaller than the sampling period T. It can be assumed in consequence that the broadband noise components in the different time slots and also the sampled-and-held noise are decorrelated and hence their contributions can be evaluated separately. The SC integrator's total output noise spectral density $S'_n(\omega)$ can be determined by means of the three subcircuits shown in Fig. 5, namely

$$S'_n(\omega) = S_{bb1}(\omega) + S_{bb2}(\omega) + S_{sh}(\omega T). \tag{2}$$

The broad-band noise components can be calculated in a straightforward manner and are given

$$S_{bb1}(\omega) = \frac{\Delta}{T} \cdot \frac{k\theta}{\alpha C} \cdot \frac{\frac{1}{\omega_{on}}\alpha^2 + \frac{1}{\omega_{eq}} \cdot \left[(1+\alpha)^2 + \left(\frac{\omega}{\omega_u}\right)^2\left(\frac{\omega_u}{2\omega_{on}}\right)^2\right]}{1 + \left(\frac{\omega}{\omega_u}\right)^2\left[(1+\alpha)^2 + \alpha\left(\frac{\omega_u}{\omega_{on}}\right) + \left(\frac{\omega_u}{2\omega_{on}}\right)^2\right] + \left(\frac{\omega}{\omega_u}\right)^4\left(\frac{\omega_u}{2\omega_{on}}\right)^2} \tag{3}$$

$$S_{bb2}(\omega) = \frac{T-\Delta}{T} \cdot \frac{k\theta}{\alpha C} \cdot \frac{\frac{1}{\omega_{eq}}}{1 + \left(\frac{\omega}{\omega_u}\right)^2} \tag{4}$$

with $\omega_{on} = (2R_{on2}\alpha C)^{-1}$, $\omega_{eq} = (2R_{eq}\alpha C)^{-1}$, and $A_0 \gg 1$.

While the derivation of both continuous-time subcircuits is straightforward, the network shown in Fig. 5(c) needs some explanation: it can be depicted of Fig. 4(b) that every time a phase Φ is active, a conductive path is created by the corresponding MOS switch transistor, and the spectra of the white noise sources present in this path are low-pass filtered by the corresponding RC time constants and/or by the op amp rolloff. The resulting circulating current is creating a noise voltage across the capacitor αC. At the moment of the switch transistor's transition from the conducting to the blocked state, the instantaneous noise voltage is "frozen" on the capacitor and transferred later, as a sampled-data signal, to the output. The resulting output noise spectrum $S_{sh}(\omega T)$ is hence periodic and can be evaluated with sampled-data techniques. The sampled-and-held noise S_{sh} can thus be considered as the output of the transientless SC network shown in Fig. 5(c) with three internal noise sources (S_1, S_2, S_3) of band-limited spectra. Note that a nonideal charge transfer is merely due to the finite low-frequency op amp gain A_0. The above-mentioned band-limitation of the white noise source outputs is accomplished here by the two-ports of Fig. 6. They represent between their input terminals the original SC network "seen" by the white noise source in the corresponding phases. A linear voltage source at the output, controlled by the resulting band-limited noise voltage on the capacitor αC, represents the continuous-time broad-band contribution $S'_j(\omega)$ (for $j = 1, 2, 3$) of the corresponding noise source $S_j(\omega)$ on the capacitor, wherefrom it is transferred as a sampled-data signal to the output. Thus the output noise spectrum $S_{sh}(\omega T)$ is the sum of all sampled broad-band spectra $S_{jh}(\omega T)$ multiplied by the z-transform transfer function (from capacitor αC to the output) evaluated on the unit circle. As it has been stated above, the inherent condition of total charge transfer in SC networks results in an undersampling of the different broad-band noise components. In order to evaluate $S_{jh}(\omega T)$, the spectrum of the continuous-time noise component $S_j'(\omega)$ has to be convoluted by a Dirac comb whose period represents a small fraction of the noise bandwidth. Because of this rather uncommon procedure, the next section has been devoted to highlight the undersampling effect.

III. UNDERSAMPLED BROAD-BAND NOISE SPECTRUM

The results presented in this section are taken from [8], where they have been derived in an extensive form. We limit us here to recall the essential steps. Let assume that $r_n(\tau)$ is the autocorrelation function of a continuous-time, stationary random signal of finite energy. By sampling this signal with period T, its autocorrelation function at time-lag $\tau = kT$ yields

$$r_{ns}(k) = \frac{1}{T} \sum_{n=-\infty}^{\infty} \delta(k-n) \cdot r_n(kT). \quad (5)$$

The spectral density of the sampled random signal is given by the Fourier-transformed of (5), namely

$$FT\{r_{ns}(k)\} = \frac{1}{T} \sum_{n=-\infty}^{\infty} r_n(nT) \cdot e^{-j\omega nT} = S_{ns}(\omega T) \quad (6)$$

and is identical to the bilateral z-transform of $r_n(nT)$ evaluated on the unit circle. Let be $r_n(\tau)$ now the autocorrelation function of a first-order low-pass filtered noise with spectral density S_{n0} at the origin and with cutoff frequency ω_c. Then

$$r_n(\tau) = (\omega_c/2) \cdot S_{n0} \cdot \exp(-|\tau|\omega_c). \quad (7)$$

By introducing this expression into (6) and with the supplementary assumption that the noise signal is sampled and held with duty-cycle Δ/T, we get

$$S_{ns}(\omega T) = \omega_c \frac{T}{2} \cdot S_{n0} \cdot \frac{\sinh(\omega_c T)}{\cosh(\omega_c T) - \cos(\omega T)}$$
$$\cdot (\Delta/T)^2 \operatorname{sinc}^2(\Delta\omega/2) \quad (8)$$

the spectrum of the sampled first-order low-pass filtered noise. In what follows, the value $\omega_c T$ will be called ratio of undersampling. Note that equation (8) holds for all ratios of undersampling (i.e., for the over- and the undersampled case). For $\omega_c T \gg 1$, the effect of undersampling appears clearly: the spectral density of the sampled noise at the origin is increased with respect to the continuous-time noise spectrum by the ratio of undersampling. In fact, the aliasing of the periodical spectrum occurs due to spectral components at the multiples of the sampling frequency in the initial spectrum. This is the explication why a supplementary flicker component is submerged by the aliased broad-band noise if a sufficiently strong undersampling can be assumed. Furthermore, for already low ratios of undersampling the power density of the sampled noise becomes uniform [8]. Hence for $\omega_c T > \pi$, the sampled noise spectrum is approximately white.

IV. COMPARISON OF THEORETICAL RESULTS AND MEASUREMENTS

A. Theoretical Results

In the foregoing two sections, all the elements have been treaten which enable the derivation of the output noise generated in a SC integrator in a closed analytical form. As it has been stated in Section II, this output noise is composed of two continuous-time broad-band components whose spectra are given by (3) and (4) and of a sampled-and-held noise contribution whose spectral density will now be evaluated.

First, the ratios of undersampling for the different broad-band noise contributions sampled on the capacitor αC have to be derived. It is evident from inspection of Fig. 6 that we are just in the case (a) in presence of a first-order low-pass filtered white noise. Hence only in this case the equations derived in Section III can be applied in a strict sense. Assuming however the cutoff frequency, formed by the RC time constants, to be several times higher than the op amp's bandwidth, the spectra in the cases (b) and (c) can be considered as approximately first-order low-pass filtered. This assumption is realistic, since in SC design the general trend goes in diminishing the capacitor values (thus the RC time constants) in order to save chip area and to achieve low-power consumption, and hence the frequency limiting element is the MOS op amp. With the above assumptions, the cutoff frequencies are

$$\omega_{ca} = (R_{on1}\alpha C)^{-1}$$
$$\omega_{cb} = \omega_{cc}$$
$$= \left(2R_{on2}\alpha C\sqrt{1/4 + \alpha(\omega_{on}/\omega_u) + (\alpha+1)^2(\omega_{on}/\omega_u)^2}\right)^{-1}. \quad (9)$$

The square of the z-transform transfer function from capacitor αC to output evaluated on the unit circle yields (for ideal op amp)

$$|H(\omega T)|^2 = \left(\frac{\alpha}{2} \frac{1}{\sin(\omega T/2)}\right)^2. \quad (10)$$

Finally, the sampled broad-band noise spectrum can be considered as white, since for all realistic cases a sufficiently strong undersampling can be assumed. The sampled-and-held noise at the output of the SC integrator is then the sum of the three white noise spectral densities multiplied by the different ratios of undersampling, weighted by (10) and by a $\sin x/x$ due to the hold function

$$S_{sh}(\omega T) = \frac{T}{2} \cdot \left[\omega_{ca} \cdot 2k\theta R_{on1} + \omega_{cb} \cdot 2k\theta(R_{on2} + R_{eq})\right] \cdot |H(\omega T)|^2 \cdot \operatorname{sinc}^2(\omega T/2)$$

$$= T\frac{k\theta}{\alpha C} \cdot \left[1 + \frac{1 + R_{eq}/R_{on2}}{2\sqrt{1/4 + \alpha(\omega_{on}/\omega_u) + (\alpha+1)^2 \cdot (\omega_{on}/\omega_u)^2}}\right] \cdot (\alpha/\omega T)^2. \quad (11)$$

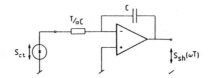

Fig. 7. Continuous-time noise model for the sampled-and-held noise contribution generated in the SC integrator of Fig. 1.

TABLE I
VALUES AND CHARACTERISTICS OF THE COMPONENTS USED FOR THE BREADBOARD REALIZATION

α	=	1
C	=	10 pF
T	=	0.1 msec
Δ/T	=	0.25
ω_u	=	$2\Pi \cdot 700$ kHz
Θ	=	293 °K
R_{on2}	=	3.5 kΩ
R_{eq}	=	1.55 MΩ

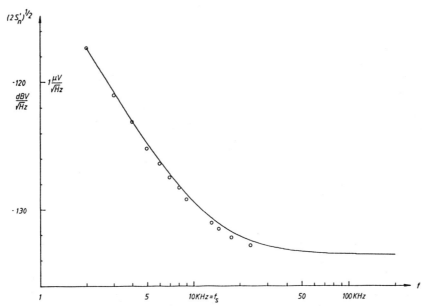

Fig. 8. Noise spectrum of the switched capacitor integrator shown in Fig. 1. —— calculated. ○ measured.

The expression (11) can be considered as the output noise of an ideal (noiseless) continuous-time integrator with T/α time constant and a white noise source S_{ct} at the input (Fig. 7), where

$$S_{ct} = T \frac{k\theta}{\alpha C}$$
$$\cdot \left[1 + \frac{1 + R_{eq}/R_{on2}}{2\sqrt{1/4 + \alpha(\omega_{on}/\omega_u) + (\alpha+1)^2 \cdot (\omega_{on}/\omega_u)^2}} \right]. \quad (12)$$

Note that the integrator's resistor value is identical to the low-frequency ($\omega T \ll 1$) equivalent of the switched capacitor αC [9]. Comparing the broad-band noise spectra (3) and (4) with (11), it is evident that the sampled-and-held noise contribution is predominating in the interesting frequency range up to the Nyquist rate, since $R_{on2}\alpha C$, $R_{eq}\alpha C \ll T$. However, if the integrator's output is sampled, the broad-band noise component has to be considered as well.

B. Measurements on a Laboratory Model

In order to verify the noise calculations presented above, the SC integrator shown in Fig. 1 has been implemented using discrete elements. This procedure allows on the one hand to determine the characteristics of the components (switch resistances, op amp bandwidth and noise) in an easy manner separately. On the other hand, the clock feedthrough could be diminuated in this way by capacitive compensation circuits and hence the instrument sensitivities increased. The component values and characteristics are shown in Table I. The plot of $\sqrt{2S'_n}$ versus frequency (solid line) with the above values is shown in Fig. 8. Excellent correlation with the measurements made on the laboratory model can be observed in spite of the neglected op amp's flicker noise component and the small ω_{on}/ω_u ratio. Note that the switch noise contribution is 3 percent of the total output noise.

V. Conclusion

In this section, some general statements on SC noise analysis will be established on base of the specific results derived in the foregoing chapters.

Noise generated in a SC network has two origins: the thermal fluctuations associated with the channel-resistances of the MOS switch transistors and the noise created in the op amp. In a first step, two assumptions have been made.

1. Since the off resistances of the MOS switches present in conjunction with the capacitors extremely great time constants, the resulting "off noise" is considered as a slowly varying offset voltage and is hence neglected in the noise spectrum. In consequence, the noise equivalent of each MOS switch transistor consists of a white noise source, a noiseless resistor (both corresponding to the on resistance of the MOST) and an ideal switch. The initial four-phase noise equivalent of the SC circuit can in this way be reduced to a resistive two-phase SC network with internal noise sources.

2. By neglecting at the same time the op amp's flicker noise component and the noise of the output stage, all the internal noise sources can be considered as white and the op amp is assumed to have a rolloff characteristic. This assumption is justified in case that the noise corner frequency ω_{cr} (see Fig. 2(a)) is smaller than the sampling rate. In this case, the flicker noise is "submerged" by the aliased broad-band noise of the op amp (cutoff frequency approximately at the op amp's unity-gain bandwidth) and ω_{cr} shifted towards lower frequencies.

The noise at the output of the SC network consists of a continuous-time broad-band component and of a sampled-and-held noise contribution whose spectral power is concentrated in the range up to the sampling rate. The broad-band component is due to all noise sources which are coupled directly to the output in at least one phase. Since the noise sources are independent, their contribution can be evaluated separately. The cutoff frequencies of the different contributions are determined by the transients ($R_{on}C$ time constants and op amp rolloff) of the SC network valid in the corresponding phases. In order to enable total charge transfer, these transients have to be much smaller than the clock period. A first consequence is that the broad-band components in adjacent time slots can be considered decorrelated, their noise contributions can simply be added. With these assumptions, the broad-band output noise spectrum of SC networks can be evaluated by classical continuous-time analysis methods. At the moment of a MOS switch transistor's transition from the conducting to the blocked state, the instantaneous broad-band noise voltage is "frozen" on the capacitor which is in parallel to the noise source, then transferred as a sampled-data signal to the output. The SC network's output noise spectral density due to this sampled-and-held noise contribution is in consequence the spectrum of the sampled broad-band noise present on the above mentioned capacitor multiplied by the z-transform transfer function from the capacitor to the output. The condition of total charge transfer implies on one hand that the sampled-and-held noise can be considered as approximately decorrelated from the broad-band noise which it is generated from. The total noise spectral density at the output of the SC network is hence approximatively the sum of the broad-band noise spectrum and the sampled-and-held noise spectral density. The total charge transfer condition in SC networks results on the other hand in the fact that the broad-band noise sources generating the sampled-and-held noise contribution are strongly undersampled. The spectrum of a sampled first-order low-pass filtered noise has been evaluated and two important results have been established in the case of undersampling.

1) The initial continuous-time broad-band noise spectrum is increased by the factor of undersampling due to aliased broadband noise component.

2) For relatively low ratios of undersampling, the spectrum of the sampled broad-band noise becomes approximately white.

As a result of the first statement, the sampled-and-held noise is predominating in the baseband of the SC networks. The second statement implies that the shape of the sampled-and-held noise spectrum is determined by the z-transform transfer function from the capacitor in parallel with the noise source to the output, weighted by a $\sin x/x$ due to the hold function. As a result, the predominating sampled-and-held noise can be determined by SC analysis programs which do not take into account the transients in SC networks, since their band-limiting role is implicitly content in the ratio of undersampling.

As a conclusion, for SC low noise design, the following points have to be considered: (a) for a given sampling rate, the bandwidth of the op amps has to be reduced to a minimum in order to prevent excessive undersampling; (b) in either of the phases, no small valued capacitors should be present in a path which is not lowpass filtered by the rolloff characteristic of an op amp.

Acknowledgment

The authors are indebted to Prof. R. Dessoulavy for his encouragement and to P. Vaucher for his help in preparing this paper.

References

[1] D. J. Allstot, "MOS switched capacitor ladder filters," Electron. Res. Lab. Univ. California, Berkeley, CA, Memo. UCB/ERL M79/30, May 1979.
[2] C. -A. Gobet and A. Knob, "Noise generated in switched capacitor networks," Inst. Elec. Eng. Electron. Lett., vol. 16, no. 19, pp. 734–735, Sept. 1980.

[3] C.-A. Gobet and A. Knob, "Noise analysis of switched capacitor networks," in *Proc. IEEE Int. Symp. Circuits Syst.*, (Chicago, IL), Apr. 1981, pp. 856–859.
[4] B. Furrer and W. Guggenbuehl, "Noise analysis of sampled-data circuits," in *Proc. IEEE Int. Symp. Circuits Syst.*, (Chicago, IL), Apr. 1981, pp. 860–863.
[5] F. Maloberti, F. Montecchi, and V. Svelto, "Noise and gain in a SC integrator with real operational amplifier," *Alta Frequenza*, no. 1, vol. L, pp. 4–11, 1981.
[6] M. L. Liou and Y.-L. Kuo, "Exact analysis of switched capacitor circuits with arbitrary inputs," *IEEE Trans. Circuits Syst.*, vol. CAS-26, pp. 213–223, Apr. 1979.
[7] J. Vandewalle, H.-J. De Man, and J. Rabaey, "Time, frequency, and z-domain modified nodal analysis of switched-capacitors networks," *IEEE Trans. Circuits Syst.*, vol. CAS-28, pp. 186–195, Mar. 1981.
[8] C.-A. Gobet, "Spectral distribution of a sampled first order lowpass filtered white noise," *Inst. Elec. Eng. Electron. Lett.*, no. 19, vol. 17, pp. 720–721, Sept. 1981.
[9] A. Knob and R. Dessoulavy, "Analysis of switched capacitor networks in the frequency domain using continuous-time two-port equivalents," *IEEE Trans. Circuits Syst.*, vol. CAS-28, pp. 947–953, Oct. 1981.

Author Index

B

Bertails, J-C., 42
Branin, F. H., Jr., 55
Bruton, L. T., 63, 70

E

Erdi, G., 33

F

Fischer, J. H., 85
Fukui, H., 13

G

Gobet, C-A., 96

H

Hartmann, K., 3
Haslett, J. W., 79
Hillbrand, H., 59

K

Knob, A., 96

M

Meyer, R., 45

N

Nagel, L., 45
Netzer, Y., 19

R

Rohrer, R., 45
Russer, P. H., 59

T

Treleaven, D. H., 63, 70
Trofimenkoff, F. N., 63, 70

W

Weber, L., 45

Editor's Biography

Madhu Sudan Gupta (S'68-M'72-SM'78) received the Master's and Ph.D. degrees in 1968 and 1972, respectively, from the University of Michigan, Ann Arbor, where he was a member of the Electron Physics Laboratory, and carried out research on the large-signal and noise characteristics of microwave semiconductor devices. During 1973-79, he was first an Assistant Professor, and then an Associate Professor of Electrical Engineering at Massachusetts Institute of Technology, Cambridge, where he was a member of the Research Laboratory of Electronics, and conducted research on microwave and millimeter wave semiconductor devices, related high-field carrier transport properties, and thermal fluctuations. Since 1979, he has been at the University of Illinois at Chicago, where he is presently a Professor of Electrical Engineering and works on the materials, design, characteristics, limitations, circuits, and noise of electron devices which are high-frequency, nonlinear, active, and/or very small. During 1985-86, he was a Visiting Professor of Electrical and Computer Engineering at the University of California, Santa Barbara.

Dr. Gupta is a member of Eta Kappa Nu, Sigma Xi, and Phi Kappa Phi; is a Registered Professional Engineer; and has served as the Chairman of the Boston and Chicago chapters of the IEEE Microwave Theory and Techniques Society. He is the editor of *Electrical Noise: Fundamentals and Sources* (IEEE Press, 1977), and was the recipient of a Lilly Foundation Fellowship.